THE ORCHESTRA IN ENGLAND

THE ORCHESTRA IN ENGLAND

A Social History

by

REGINALD NETTEL

1948
READERS UNION/JONATHAN CAPE
LONDON

Republished 1972
Scholarly Press, Inc., 22929 Industrial Drive East
St. Clair Shores, Michigan 48080

Library of Congress Cataloging in Publication Data

Nettel, Reginald, 1899-
    The orchestra in England.

    Reprint of the 1948 ed.
    Bibliography: p.
    1. Orchestra. 2. Music--England--History and
criticism. I. Title.
ML1231.N48 1972   784'.06'610942   75-181219
ISBN 0-403-01630-4

NOV 25 '74

This volume was produced in 1948 *in complete conformity with the authorized economy standards. First published in* 1946 *by Jonathan Cape Ltd., it is set in Monotype Baskerville* 11 *on* 12 *point and was reprinted at the Alden Press in the City of Oxford. It is one of the books produced by Readers Union Ltd., of* 38 *William IV Street, Charing Cross, London, and of Letchworth, Hertfordshire, for sale to its members only. Membership of RU may be made at all bookshops and particulars are obtainable at the above addresses*

CONTENTS

BOOK ONE
GROWTH OF THE CLASSICAL ORCHESTRA

INTRODUCTION	17
THE PRE-CLASSICAL ORCHESTRA	29
HANDEL'S CONTRIBUTION	44
THE ENGLISH MANNER	61
SUBSCRIPTION CONCERTS	73
HAYDN IN LONDON	84

BOOK TWO
THE PHILHARMONIC PERIOD

THE FOUNDING OF THE PHILHARMONIC SOCIETY OF LONDON	103
TEMPORA ET MORES	112
ENGLISH RELATIONS WITH BEETHOVEN	120
WHY WEBER CAME TO LONDON	142
THE MENDELSSOHN TRADITION	147
THE ORCHESTRA IN LOW LIFE	159
COSTA AND BERLIOZ	171
WAGNERIAN INTERLUDE	183
SIR CHARLES HALLÉ	190

BOOK THREE
THE NATIONALIST PERIOD

MANNS AND GROVE AT THE CRYSTAL PALACE	203
HANS RICHTER	217
MUNICIPAL MUSIC UNDER SIR DAN GODFREY	230
THE TWENTIETH CENTURY ORCHESTRAS	243
REFORMATION	269
INDEX	293

ILLUSTRATIONS

VIEW OF THE ORCHESTRA AND PERFORMERS IN WESTMINSTER ABBEY DURING THE COMMEMORATION OF HANDEL. 1784 *facing p.*	64
A CONCERT TICKET BY HOGARTH	65
PLAN OF THE ORCHESTRA AT THE HANDEL COMMEMORATION	96
THE B.B.C. SYMPHONY ORCHESTRA, 1942	97
A PROMENADE CONCERT — JULLIEN. From *Punch*	164
PLAN OF THE PHILHARMONIC ORCHESTRA UNDER COSTA	174
GRAPH SHOWING NATIONAL ORIGIN OF NEW COMPOSITIONS PLAYED BY THE PHILHARMONIC SOCIETY OF LONDON. 1823 to 1912	259

To
MY MOTHER

PREFACE

WITHIN memory of the present generation the orchestra has come to take an important place in our daily lives; famous conductors have replaced prima donnas in general esteem, and choral works — all-important thirty years ago — now take second place to symphonies. Many factors have gone to make this change in taste, and these are social history, yet no book dealing specifically with the history of the orchestra in England has been available. It is hoped therefore that this book will prove useful to all lovers of orchestral music.

It must be remarked, however, that this is a history and not a textbook on orchestration. There are so many good books dealing with the nature of orchestral instruments and the scoring of orchestral works that to add yet another to their number would be to court redundancy. Such would certainly be the case with the modern orchestra, but it is felt that the pre-classical orchestra differed so much from the modern orchestra that some explanation of its nature would not be unwelcome. The first five chapters therefore devote some little space to a consideration of the instruments and manner of performance of orchestras prior to Haydn and Mozart. The reader who is interested only in the orchestra he sees in the modern concert-hall may skip these chapters if he so desires, but it is felt nevertheless that his appreciation of pre-classical works as they are played to-day will be improved by a knowledge of the orchestra as it existed before Haydn.

Much that is familiar to musical historians naturally reappears in this book, but it has nevertheless been thought desirable to give references to sources fairly consistently because the book is intended mainly for the general reader. My thanks are due especially to the following for permission to quote copyright material. Sir Thomas Beecham and Messrs. Hutchinson, Ltd., for matter from *A Mingled Chime*; the same publishers, and Mr.

PREFACE

Thomas Russell for the use of the latter's *Philharmonic Decade*; Messrs. Methuen and the Exors. of the late Joseph Bennett, for *Forty Years of Music*; Messrs. Secker and Warburg and Mr. Frank Howes, for *Full Orchestra*; Lady Jessie Wood and Mr. Victor Gollancz for the late Sir Henry's *My Life of Music*; *The Times* newspaper and *The Manchester Guardian*; the officials and staff of the Bodleian Library for much assistance, unfailing courtesy, and permission to reproduce all illustrations used in this book except that facing page 97, for which I am indebted to the British Broadcasting Corporation.

A short bibliography is included, showing books that have been consulted, and from which the reader who desires more detailed information may profit.

R. N.

1945

The need for a second reprinting gives me the opportunity to thank all who have drawn my attention to errors and omissions in the first edition. These I have endeavoured to correct and add.

R. N.

1947

BIBLIOGRAPHY

ANDERTON, H. O., *Granville Bantock*, John Lane, 1915.
BEECHAM, SIR THOMAS, *A Mingled Chime*, Hutchinson, 1944.
Beethoven's Letters, trans. by Lady Wallace, Longmans Green, 1866.
BENAS, BERTRAM B., 'Merseyside Orchestras': Transactions of the Historic Society of Lancashire and Cheshire, vol. 95.
BENNETT, JOSEPH, *Forty Years of Music*, Methuen, 1908.
BURNEY, DR. CHARLES, *General History of Music*, 1776-89.
BURNEY, DR. CHARLES, *An Account of the Musical Performances in Westminster Abbey and the Pantheon in 1784 in Commemoration of Handel.*
CARSE, ADAM, *The Orchestra in the Eighteenth Century*, Heffer, 1940.
CHORLEY, H. F., *Thirty Years' Musical Recollections*, 1862.
FOSTER, M. B., *History of the Philharmonic Society of London*, John Lane, 1913.
GODFREY, SIR DAN, *Memories and Music*, 1924.
HADDON, J. C., *Haydn*, Dent.
HALLÉ, SIR CHARLES, *Life and Letters*, edited by C. E. and Marie Hallé.
HAWEIS, REV. H. R., *Music and Morals*, 1871.
HAWKINS, SIR JOHN, *General History of Music*, 1776.
HOGARTH, GEORGE, *History of the Philharmonic Society of London*, 1862.
HOGARTH, GEORGE, *Musical History, Biography, and Criticism*, 1835.
HOWES, FRANK, *Full Orchestra*, Secker and Warburg, 1942.
Mendelssohn's Letters from Italy and Switzerland, trans. by Lady Wallace, Longmans Green, 1864.
PEPYS, SAMUEL, *Diary*.
RUSSELL, THOMAS, *Philharmonic Decade*, Hutchinson, 1945.
SHORE, BERNARD, *The Orchestra Speaks*.

BIBLIOGRAPHY

STREATFIELD, R. A., *Handel*, Methuen, 1909.
TERRY, CHARLES SANFORD, *J. C. Bach*, Oxford, 1929.
TURBERVILLE, A. S., *English Men and Manners in the Eighteenth Century*, Oxford, 1926.
WOOD, SIR HENRY J., *My Life of Music*, Gollancz, 1938.
YOUNG, G. M., *Early Victorian England*, Oxford, 1934.

BOOK ONE

GROWTH OF
THE CLASSICAL ORCHESTRA

INTRODUCTION

IT is a paradox of history that one of the greatest seafaring nations in the world should have earned for itself a reputation for insularity. Since Great Britain is an island, it would be futile to deny that before the modern craze for speedy and cheap transport came, the great majority of non-seafaring Britishers had little opportunity for studying foreign styles of music in their lands of origin; such music had to be brought to Britain, and the difficulties attendant on this course had to be overcome. The result of this was that such music became the prerogative in this island of those who could afford to pay high prices: not until the twentieth century did it become possible for foreign orchestras like the Berlin, Vienna or Prague Philharmonic economically to tour the British Isles. Hence the paradox — for while much money was being spent in the attraction of foreign artists to this country, the great mass of the lower and lower-middle classes were unfamiliar with the orchestra and its music, and could give little assistance towards its establishment. The story of the symphony orchestra in England is, therefore, a story of foreign instrumentalists visiting our land and striving to establish their art under the peculiar economic and social conditions obtaining here, while British taste, striving at first to resist their influence, failed utterly, and only after two hundred years of foreign domination came to understand that within the framework of this foreign style of music it was possible to make a distinctive contribution, forming as useful an indication of our national character as the effects of our seafaring and colonization do in the scheme of world affairs.

In order to understand this it is desirable to start at the time when British colonization started. We were not at that time insular in our musical taste. In the same year that the Spanish Armada was defeated and William Byrd produced his *Psalms, Sonnets and Songs of Sadness and Piety*, Nicholas Younge brought out a collection of continental madrigals under the title of *Musica Transalpina*. Nor was music the prerogative of those who lived near cathedrals, as Younge did; it was to be found in the homes, and even on the high seas, for there is in existence a report sent in 1578 by Captain San Juan de Anton to the Spanish government, describing his defeat by Sir Francis Drake off the Ecuador coast in that year, and how he was taken wounded aboard *The Pelican*. 'On that ship was much singing', he reported, and 'Drake dined alone, with music'. A very different state of affairs from the early nineteenth century when our music was in the hands of foreigners.

It is a matter of regret that we failed during the Commonwealth to produce any musician worthy of comparison as an artist with Milton, but that is only indicative of the truism that while the struggles of an experimental age may feed the inspiration of a great artist, they have no power to influence nature in a way that will engender men of genius. Such composers as we had during the Commonwealth received their due meed of praise — indeed, Milton's sonnet on Henry Lawes goes too far when it says:

> Henry whose tuneful and well measured song
> First taught our English music how to span
> Words with just note and accent,

and neither Lawes nor any of his contemporaries have left anything of the intellectual and emotional value of *Paradise Lost*, the standard by which greatness in the Commonwealth period must be judged.

If there was not greatness, however, there was a continuation of the Tudor tradition of music in the home. Anthony Wood of Oxford wrote often in his autobiography of weekly musical meetings at the home of the organist of St. John's College, 'situate and being in a house, opposite to that place whereon the Theatre was built'. There he would meet:

> Will Ellis, Batchelor of Musick, and owner of the house, who always played his part either on the organ or virginal:

INTRODUCTION

Dr. John Wilson, the public professor, the best at the lute in all England. He sometimes played on the lute, but mostly presided[1] the consort. — Curteys, a lutenist, lately ejected from some choir or cathedral church. Thomas Jackson, a bass violist ... Ed. Low, Organist lately of Christ Church. He played only on the organ; so when he played on that instrument Mr. Ellis would take up the counter-tenor viol, if any person were wanting to perform that part. Gervace Littleton ... a violist; he was afterwards a singing man of St. John's Coll. Will Glexney, who had belonged to a choire before the warr ... he played well upon the bass viol, and sometimes sang his part ... Proctor, a young man and a new comer. John Parker, one of the university musitians. But Mr. Low, a proud man, could not endure any common musitian to come to the meeting, much less to play among them. Of this kind I must rank John Haselwood, an apothecary, a starch'd formal clister-pipe, who usually played on the bass viol, and sometimes on the counter-tenor. He was very conceited of his skill (though he had little of it) and therefore would be ever and anon ready to take up a viol before his betters, which being observed by all, they usually called him 'Handlewood'. The rest were but beginners.

The link with the Tudor tradition is clear enough in the fact that Will Glexney would either play or sing his part, and that the conservative Mr. Low could not endure to have a common musician come among them, much less play in consort with them. The Elizabethan belief that the ability to sing or play in parts was a mark of good breeding was beginning to lose force, however, for 'Handlewood' came to the meetings and played after his fashion, and we can see this social supremacy of part-playing declining still further as a new instrument called the violin began to receive attention. As can be expected, it did not come among them without some measure of conservative criticism, for

    A.W. was now advised to entertain one **Will James**, a

[1] Directed. See page 39.

GROWTH OF CLASSICAL ORCHESTRA

dancing master, to instruct him on the violin, who by some, was accounted excellent on that instrument, and rather, because it was said that he had obtained his knowledge in dancing and musick in France. He spent, in all, half a yeare with him, and gained improvement from him; yet at length he found him not a complete master of his facultie, as Griffith and Parker were not; and, to say the truth, there was no complete master in Oxon for that instrument, because it had not hitherto been used in consort among gentlemen, only common musitians, who played but two parts. The gentlemen in private meetings, which A.W. frequented, played three, four and five parts with viols, as treble viol, tenor, counter-tenor, and bass, with an organ, virginal or harpsicon joyn'd with them; and they esteemed a violin to be an instrument only belonging to a common fiddler, and could not endure that it should come among them, for feare of making their meetings to be vaine and fiddling. But before the restoration of King Charles II. and especially after, viols began to be out of fashion, and only violins used, as treble-violin, tenor, and bass violin; and the King, according to the French mode, would have 24 violins played before him while he was at meales, as being more airie and brisk than viols.

Charles set himself completely to reform the ultra-conservative musical taste of the English gentlemen. That he was destroying an English musical tradition that had borne us much honour since the days of Henry VIII did not apparently concern him. In his determination to establish French styles in England Charles spared no effort, and in those departments where he had control — as in the Chapel Royal — he had his way; but his subjects had wills of their own — they respected the King's wishes, but they expected him also to respect theirs. Blood had been shed on that issue not so long ago, and the least said about it the better, but if the English regretted their recent lapse into regicide they were not prepared to tolerate even now a royal control of conscience, opinion, or taste.

Charles' efforts to modernize English taste by bringing it in

INTRODUCTION

line with the fashion of the court of Louis XIV induced, therefore, some searching of personal opinions among his subjects. The Merry Monarch, some of them thought, might easily be persuaded like Sir Andrew Aguecheek to 'go to church in a galliard and come home in a coranto'. Samuel Pepys had an ear for most of the gossip about town, and ideas of his own about music. He saw Pelham Humphreys after his return from training in the court of Louis XIV, and formed an opinion of him at once characteristically Pepysian and English:

> Thence I away home... and there find, as I expected, Mr. Caesar and little Pelham Humphreys, lately returned from France, and is an absolute Monsieur, as full of form and confidence, and vanity, and disparages everything and everybody's skill but his own... and after dinner we did play, he on the theorbo, Mr. Caesar on his French lute, and I on the viol, but made but mean musique, nor do I see that this Frenchman do so much wonders on the theorbo, but without question he is a good musician, but his vanity do offend me.

Pepys indeed had more sympathy for John Banister, who, like Humphreys, had been sent for training to the French court. Banister was the son of one of the waits of St. Giles-in-the-Fields, and learned his early fiddling under his father. After his return from training in France he was appointed leader of the king's band of twenty-four violins, but unlike Pelham Humphreys he did not return 'an absolute monsieur'; he was dismissed for saying in the king's hearing that he thought English violinists were superior to French, and a Frenchman named Louis Grabu was appointed in his stead. Pepys records:

> They talk how the King's violin Banister is mad. That a Frenchman is come to be some part of the King's music.

And he went in due course to form an opinion on Grabu, which he gives in his diary under the date October 1st, 1667:

> To White Hall; and there in the Boarded Gallery did hear

the musick with which the King is presented this night by Monsieur Grebus, the master of his musick; both instrumental — I think twenty-four violins — and vocall; an English song about peace. But, God forgive me! I never was so little pleased with a consort of musick in my life. The manner of setting the words and repeating them out of order and that with a number of voices, makes me sick, the whole design of vocall musick being lost by it. Here was a great press of people, but I did not see many pleased with it, only the instrumental musick he had brought by practice to be very just.

Pepys was a follower of the style of Henry Lawes, who, in setting his words 'with just note and accent', developed a declamatory style copied in Pepys' own efforts at composition, and although Pepys approved the performance of the twenty-four violins, and a truly great English composer, Henry Purcell, came to write music for the court of St. James that suited both the king and the musical public, yet the fashion for private orchestras, private opera houses, and domestic composers, that had by this time become general on the continent, did not find general acceptance here.

Many things contributed to this. Neither Pelham Humphreys nor Henry Purcell lived long enough to establish their art on a permanent basis among the conservative amateurs by whom they were surrounded, but most of all, the lingering feudalism that made possible the economic structure of France no longer had any hope of revival in England, where Parliament controlled the means of taxation. Without the social system prevalent among the spendthrift autocrats of the continent, the development of the orchestra in England had to proceed along different lines, depending in fact not so much upon single individuals as upon groups of professional musicians seeking to form semi-private clubs with the assistance of influential amateurs, or upon capitalist enterprise.

John Banister led the way. On his dismissal from leading the king's band, he set up as a music teacher, but with an interesting

INTRODUCTION

and original sideline to that retiring business, for a notice in the *London Gazette* on December 30th, 1672, stated:

> These are to give notice that at Mr. John Banister's House (now called the Music School) over against the George Tavern in Whyte Fryers, this present Monday, will be musick performed by excellent Masters, beginning at four of the clock in the afternoon, and every afternoon for the future precisely at the same hour.

Banister had decided to commercialize the kind of musical gathering to which Anthony Wood refers in his autobiography. By so doing, he started a fashion that was to grow into our modern conception of concert promotion — a system under which the public is admitted on payment of a fee to hear music performed. Such an idea had not previously been put into practice, and it did not appear on the continent until the eighteenth century.[1] Banister's experiment met with success, and in consequence soon found imitators. Business rivalry, however, appears not to have entered objectionably into the practice during Banister's lifetime, for he often played at the weekly concerts organized by his rival, Thomas Britton, the musical small-coals man.

Britton's career throws a most interesting light on the social influence of music, for Thomas Britton, by his love of music, was able to mix with people of all classes in a decidedly class-conscious age. He, a man who earned his livelihood by selling charcoal (which he carried on his back through the streets of London) was yet the respected friend of scholars, poets, artists, musicians, and men of rank. He helped in the formation of the Harleian Library, the Somers tracts he collected entirely himself, the portrait of him which now hangs in the National Portrait Gallery was painted by Woolaston, a regular attender at his weekly concerts, at which all the most eminent musicians of the day appeared, including, besides Banister, Dr. Pepusch,

---

[1] The Concert Spiritual of Paris started in 1725. It was not the first society to be founded on the continent for concert-giving, but Paris as usual set the fashion.

GROWTH OF CLASSICAL ORCHESTRA

Henry Needler, and Handel. Britton was respected for his conversation and knowledge of books and science, besides the good taste shown in the concerts he provided. These were held in a long narrow loft over his shop, approached by an outside stairway. At first there was no charge for admission, but later a subscription of ten shillings a year was the rule, and Britton provided coffee at a penny a cup. Notwithstanding this simple environment, Britton's concerts attracted the best performers of his day and an audience comprised entirely of men of good taste and breeding.

It would be idle to pretend that a poor man having contacts with others of the higher social classes would not come under suspicion in those days. Class distinctions were sternly enforced in the eighteenth century, as no doubt they often needed to be on evidence such as is shown in the works of Defoe and Hogarth. The common man's view of the class barrier is not evident in the instrumental music of the time, but in the folk-songs of the time, when a common theme was that telling of a youth seized by the press-gang because he had won the heart of the squire's daughter:

> My lover was a poor farmer's son
> When first my tender heart he won.
> His love to me he did reveal.
> I little thought of the *Nightingale*.
> My cruel father made it so
> That he from me was forced to go,
> A press gang sent that did not fail
> To press my love on the *Nightingale*.

Folk-songs stamped a man as a 'common musician', but concerts were different. Thomas Britton suffered no persecution for his musical activities: for although he was variously suspected of being a Jesuit, a Presbyterian, an atheist, and a magician, and his musical gatherings said to be a cloak for such activities, these things were said only by the ignorant — his honesty was never in question among those who knew him.

This is the remarkable fact upon which an understanding of

INTRODUCTION

the cultural life of the seventeenth and eighteenth centuries largely depends. Rochester's epitaph on Charles II runs:

> Here lies our sovereign lord the King,
>   Whose word no man relies on;
> Who never said a foolish thing,
>   Nor ever did a wise one.

But surely the Merry Monarch's founding of the Royal Society was an act of some wisdom? If it did not confine its membership to those able to claim some scientific reputation, at least it created a fashion for the respect of scientists and their work which the French court copied and introduced to the other courts of Europe. Whatever the political situation might be, the thinker — be he scientist, philosopher or artist — could move freely from court to court and be hospitably received, for even if his host had little capacity for thought he was conscious of the prestige that such hospitality would bring.

Besides national barriers, class barriers were in this way rendered less impregnable to the artist and the thinker. It was this fashion for culture that enabled the French philosophers of the eighteenth century to spread their doctrines, for although they themselves were of the middle classes, they were received in the houses of the nobility, and their works read by the nobility, who adopted a liberal view towards their philosophies while they continued their extravagant social and economic policy. In the British Isles our Parliamentary system dulled the edge of discontent among the middle classes, for it offered at first a means of reform, and when, in the eighteenth century, Parliament sank to its lowest depths in corruption, the Industrial Revolution came with a promise of wealth for the middle-class speculator. The result of all this was a more general tendency for ambitious men in Britain, throughout the period of unrest abroad, to have more faith in the essential goodness of the social system. In cultural circles there was elasticity of class distinctions that made the popularity of Thomas Britton' concerts possible, and which stretched so far that intermarriage

GROWTH OF CLASSICAL ORCHESTRA

was actually permitted between some musicians and ladies of quality.

The eighteenth century in England was a time, then, of considerable freedom and enterprise among musicians; there grew up many public concert-giving clubs after the style of those of Banister and Britton. Of these one of the most important was the Academy of Antient Music, which met at the Crown and Anchor in the Strand under the artistic guidance of Dr. Pepusch; its members included the Earl of Abercorn, Henry Needler and Mulso. For a time this club had a rival in one called the Philharmonic Society, founded in 1728 at the Devil Tavern at Temple Bar, organized by Dr. Maurice Greene.

The extent to which a musician was free depended greatly on his reputation with the public. Dibdin had a hard struggle at first, selling his early songs to the publishers of broadsheet ballads, whose usual offer for a good song was no more than a few shillings. After he began to write for the theatres, however, and his name became coupled with popular sea-songs, Dibdin thought quite reasonably that he should be paid a rate comparable with his value to the management, but it was not easy to obtain such terms as he wished. 'He wrote with great industry for over twenty years, during which time, according to an account given by himself, he produced nearly a hundred operas, and other musical pieces, for the different theatres. He adds, that for all these pieces, during so long a period, the whole amount of his emoluments, including the salaries for conducting the music at different theatres, and his annual benefits, was only £5,500. This very inadequate remuneration he ascribes to the unfair dealings of managers, with all of whom, and especially Garrick, he seems to have been engaged in constant quarrels. Finding himself so ill rewarded for his theatrical labours, he set on foot a series of entertainments, consisting of recitations and songs — written, composed, delivered and sung by himself.'[1]

In fact, Dibdin found that only by taking the commercial risk

[1] Hogarth, *Musical History, Biography, and Criticism.*

INTRODUCTION

himself was he able to obtain an adequate reward. From a small theatre in Leicester Square he finally went on tour in the Provinces.

Capitalist enterprise held the reins, but fortunately concert promotion was still done in a small way, so that a single musician with a good artistic reputation was able to attract audiences sufficiently large to yield him a profit. Few, however, were like Dibdin, who took all the risk himself; the general method was to gain first the support of influential patrons who would subscribe to a series of concerts. Capitalist enterprise as we now understand the term was more in evidence in the organization of the London pleasure gardens — Vauxhall Gardens and Ranelagh House and Gardens. The former had the longest period of prosperity, for it was first opened as the Spring Gardens in the seventeenth century; Samuel Pepys went there on May 28th, 1667, where he saw

> A great deal of company, and the weather and garden very pleasant: that it is very pleasant and cheap going thither, for a man may go to spend what he will, or nothing, all is one. But to hear the nightingale and the other birds, and here fiddles, and there a harp, and here a Jew's trump, and here laughing, and there fine people walking, is mightily divertising.

Such attractions could not be long unexploited. Vauxhall Gardens remained open until after 1850, reaching their peak of popularity during the Regency: their rise and ultimate decline into the lowest depths of vulgarity is in itself a lucid illustration of the trend of taste during the Industrial Revolution under purely capitalist direction, but we are here interested in Vauxhall Gardens chiefly because of their importance in the eighteenth century. Music was a popular attraction, but not the most popular, for admission to the gardens cost two shillings when concerts were given, but five shillings for firework displays.

There was an important performance in the Green Park on

GROWTH OF CLASSICAL ORCHESTRA

April 27th, 1749, when a great firework display was arranged to celebrate the Peace of Aix-la-Chapelle, with music by Handel. The final rehearsal of this *Music for the Royal Fireworks* was given in public in Vauxhall Gardens by an orchestra of a hundred players. Twelve thousand people paid for admission a fee of two shillings and sixpence each, and for three hours London Bridge was blocked with the flow of carriages. This sort of mass entertainment given for profit had possibilities not to be ignored, but it remained for the business men of the nineteenth century to make the most of it at the Crystal Palace and elsewhere. Mass musical entertainment, however, is not always a guarantee of a love of music, for during the latter half of the eighteenth century the concerts in Vauxhall Gardens started at 10 p.m., but the public developed a habit of arriving at midnight. In order to preserve the impression that people really went there to listen to the music, the proprietors advanced the time of starting the performances to 11 p.m.

Under these conditions orchestral music began in London — in the theatres, in the new concert rooms, and in the public gardens, with orchestras varying in size but rarely more than a large chamber music party (Handel's orchestra for the Firework Music was exceptional); indeed, it was many years after the eighteenth century had run its course before chamber music ceased to have its place in orchestral concerts in London. The players were free to take engagements from any of the various concert promoters, and so grew accustomed to playing at sight and adaptable to strange leaders. They could not under these conditions learn the discipline that belonged to the permanent private orchestras on the continent, nor had the composer the same facilities for the development of the new symphonic forms that he would have in daily rehearsals with a private orchestra. These facts are significant. Each system, however, had its own contribution to make to the development of the modern orchestra and its music.

THE PRE-CLASSICAL ORCHESTRA

THERE was a freedom in the musical works of the seventeenth and eighteenth centuries that is apt to cause some confusion in the mind of the twentieth-century enthusiast. Two hundred years ago names which now have a definite meaning were only being tentatively applied to orchestral compositions that were in themselves experiments in new forms of musical expression. Boyce used the word 'symphony' to describe a form of composition built on the Italian *Sinfonia avanti l'opera*, but had he chosen the title 'Overture', as J. C. Bach did for compositions very similar, nobody would have demurred; even more applicable was the term 'concerto grosso', for it was used to describe almost any piece of concerted instrumental music in the early eighteenth century, from grouped movements in dance forms to the strongly individual 'Brandenburg' concertos of J. S. Bach. The eighteenth century was an age, in fact, of freedom and initiative that was to become straitened and formalized by its last decade, and was striving throughout its course towards this end. The cultured man of the eighteenth century disliked loose nomenclature and vague meanings even more than we do: his freedom and initiative in instrumental music arose from the fact that he was grappling realistically with problems that his forebears had too long evaded. He would not conform to accepted forms of instrumental music because such forms were not completely satisfactory, being derived either from short dance tunes or vocal styles of composition.

This enforced experimentation among a class of people who admired formal perfection more generally than those of any other age induced a dissatisfaction in their minds that proved a spur to their resources of invention. We of the twentieth century have no right to assume, as we so often do, that a formal mind is necessarily reactionary. The progress made in instrumental music during the pre-classical period must not be

GROWTH OF CLASSICAL ORCHESTRA

judged by the progress made during the romantic period of the nineteenth century, for the romanticists built on a foundation of classicism. A better aid to the appreciation of the preclassical struggle will be found in a comparison of their efforts with the slow evolution of vocal polyphonic music from Hucbald to the sixteenth-century madrigalists.

Instrumental composers had to start with some very poor interpreters. It is said that Lully's players played at first only by ear, and that he had not only to evolve a new style of concerted playing, but teach them how to read their parts. We have seen in England how John Banister came not from the educated class known to Anthony Wood, but from the Waits of St. Giles-in-the-Fields. Banister had learnt the violin under his father, but there is no reason why some good instrumental players should not come from such a source, for John Ravenscroft, a composer of much admired English dance music in the early eighteenth century, was a Wait and a theatre musician. In fact, it was to these musicians, and not to men of good birth and education such as Anthony Wood describes, that the composers of the new orchestral music had to look for the rank and file of their orchestras; they had to take these players as they found them and train them to perform the new music.

The musical prowess of the town bands, then, had something to do with the orchestration of seventeenth-century compositions, but the commencement of the Waits' instrumentation was not with the violin but the oboe, or its predecessor the shawm — strings did not come into the town bands until after oboes were well established there, and strings can never have been as reliable for outdoor use in our English climate as wind instruments were. Strings had their place, of course, in the parish churches. When, therefore, orchestras came to be formed in England and Germany, one of their most marked features was their strength of oboe and bassoon tone, these instruments being almost as numerous as the strings.

Certainly the jigs and hornpipes the Waits played are nimble enough, and do not suggest that their skill was less than that of

THE PRE-CLASSICAL ORCHESTRA

the educated players, but their wind instruments were, by our standards, crude, and could not play long in tune. This applied not only to English instruments but to those in the best orchestras on the continent. It was a fault that music-lovers regretted, but for long were obliged to overlook. 'The defect, I mean, is the want of truth in the wind instruments. I know it is natural for these instruments to be out of tune', says Burney, and his rival Hawkins says of the flute that it 'still retains some degree of estimation among gentlemen whose ears are not nice enough to inform them that it is never in tune'. The eighteenth-century musician had not only to create new musical forms and train players to play them, but he had to wait for instrument makers, working in co-operation with players, to improve their instruments until the sounds the composer wished to issue from them became possible.

Strangely enough it was the simple instruments of common musicians that proved to be most suitable for use in the new orchestra — violins, flutes, oboes, bassoons, trumpets, hunting horns, and drums — and not the elaborate lutes and viols that had for so long been respected in comfortable society. The eighteenth century saw them become, if not perfect, at least capable of a satisfactory standard of performance, and with their improvement went an increasing rejection of the conservative ideas of people like Thomas Mace (who could not tolerate 'scoulding violins') and of all those who estimated part-writing of more importance than musical form.

This is one of the most interesting features of eighteenth-century music. The common musician's style of performance grew out of his knowledge of his instrument, whereas the educated musician's instrumental style was adapted to forms that grew out of sixteenth-century vocal music, as the fantasia grew from the madrigal, and fugue from the overlapping compass of voices.

As instruments became more independent, the tendency of early composers to treat them vocally proved a mixed blessing. It was this determination of composers to make brass instruments compete with voices that created the extremely difficult

GROWTH OF CLASSICAL ORCHESTRA

horn and trumpet obbligato parts to Henry Purcell's and Handel's choral and orchestral works. These composers were giants triumphing over the restrictions of their time. Of far less merit, perhaps, were the men who played trumpet fanfares at court gatherings, but it was to their styles that the composers for the new orchestras had to turn in order to discover the secret of modern orchestral tone.

We can see this recognition of the trumpet's peculiar features coming to the fore in the fifth Symphony of Boyce, where its themes are in no way akin to Boyce's string and wood-wind themes. But the trumpet's dependence on the natural scale of the harmonic series forced its peculiarities into recognition in England before the distinctive features of other instruments were generally recognized. Strings and wood-wind instruments played the same parts in numerous compositions, and at the same pitch — violins doubled by oboes and 'cellos by bassoons. Sometimes they repeat the same phrase alternately — that is, antiphonally — and this is a vocal trick borrowed from the church, but since fiddles could play a hornpipe or a rigadoon as efficiently as an oboe or flageolet, the Waits had never seen the necessity of regarding them as strangers to each other's styles: this was one of the things that had to come to light as the new orchestra developed.

The struggle for the emancipation of the wood-wind instruments occupied the major portion of the eighteenth century. It has been held by some writers that the early eighteenth-century composers had no feeling for tone-colour, because they treated all instruments so much alike. This I believe to be a mistaken view. Some composers were concerned with the maintenance of contrapuntal styles, while others were breaking with this tradition and founding a new symphonic style. The contrapuntal composers used the tone-colours of instruments to distinguish the various lines of melody — they cannot be said therefore to have ignored them; the new type of composer was at first concerned with getting his instruments to blend into some sort of harmonic *tutti* with the help of a *basso continuo*; only

THE PRE-CLASSICAL ORCHESTRA

after that was he concerned with tone-colour, but he was by no means ignorant of it. It is a common thing in early eighteenth-century scores for oboes to be marked silent for a whole movement while flutes are used. It has been suggested that the reason for this was that certain oboists would also play the flute, and so had to lay aside the oboe when flute parts were needed. But there were often many oboes and all did not double on the flute. No; it can only be assumed that the remaining oboes were silenced because the composer wanted pure flute tone.

So with clarinets. They were later in their development than the other wind instruments, but when players on the new instrument appeared, composers did not hesitate to use them. There is no evidence of eighteenth-century composers setting their minds firmly against a new instrument as Mace in the seventeenth century set his mind against the violin. The scarcity of clarinet parts in eighteenth-century music was due to the scarcity of clarinettists. It is possible that for the best part of the century no player in England could make a living out of playing the clarinet alone: he had to play the oboe, with the clarinet as a second instrument available if wanted.

Handel's opera *Tamerlano* provides a useful illustration of how clarinets would find their way into the orchestra in England. Handel's practice was to distinguish the character of the stage situation whenever possible by some associated tone-colour. Pastoral scenes therefore demanded some instrument to suggest shepherds' pipes; but not the oboe, for this was a normal part of the orchestra *tutti* of those days. In a pastoral scene in *Tamerlano*, Handel's score calls for two cornetts to accompany the air *Par che mi nascea*. Cornetts were wood-wind instruments with a cup-shaped mouthpiece such as is used for the trumpet. (The bass cornett curled in and out in a zigzag fashion and was called the serpent. This is the instrument whose tone drew from Handel, when first he heard it, the sarcastic comment, 'Aye, but not the serpent that seduced Eve'.) Cornett tone, whether it was the hard shriek of the treble or the harsh grunt of the bass, was not satisfactory, and no surprise need be

GROWTH OF CLASSICAL ORCHESTRA

felt that Handel dispensed with them when an opportunity came to employ substitutes; there is a copy of *Par che mi nascea* in the handwriting of Smith, Handel's amanuensis, where these cornetts are replaced by 'clar. et clarin. 1° et 2°'. That was in 1725, but by 1728 again no clarinets were available, for in the opera *Riccardo* Handel accompanies a pastoral air with two chalumeaux. The chalumeau was a small instrument with a cylindrical bore and a single reed, the immediate precursor of the clarinet, and apparently the next best thing available. It is possible that two clarinettists were on a visit to this country during the run of the opera *Tamerlano*, and Handel was able to employ them, but they could not have stayed long. There is no mention of clarinets again until 1762 when Arne used them in his opera *Artaxerxes*. They were again used in 1763 by J. C. Bach in his opera *Orione ossia Diana vendicata*, but the practice of using them as a second instrument by players whose main interest was in the oboe seems to have persisted. This probably held up the progress that otherwise could have been made in the use of the clarinet, for so long as oboe and clarinet parts had to be interchangeable, clarinets in C only could be used. This would seriously restrict the range of keys available for the clarinet, especially when we take into consideration the simple fingering system then in use, and the fact that all wood-wind instruments were made of boxwood, a material that stands up well to climatic changes, but is a difficult wood from which to produce good tone.

The bassoon had a place in the eighteenth-century orchestra as important as the oboe, its function being to double the bass part. It was unusual for the bassoon to take any other part, but there were exceptions. The size of the bassoon rendered the problem of intonation even more difficult on that instrument than on the smaller wood-wind instruments. The finger-holes had to be arranged where they could be covered by the fingers, and fell therefore into two groups of three. There were in addition quite early in the century extra keys for F and G sharp, B flat and D. Its lower compass even at that time was the same

THE PRE-CLASSICAL ORCHESTRA

as to-day, extending to B flat, a tone below 'cello C, and its double reed was constructed like our modern bassoon reed; the possibilities of bassoons were therefore greater than the composers of that time required, and it has, during the development of the modern orchestra, undergone less fundamental change than the other wood-wind instruments. Even the treble of the double-reed family — the oboe — has changed in an important detail: its reed was, in the eighteenth century, constructed like a bassoon reed (i.e. fairly broad at the tips), but to-day it is considerably narrowed, producing in consequence a different tone-quality. Indeed it might be desirable to distinguish between the modern and the old instrument by calling one the oboe and the other the hautboy were it not for the offence it might give to literary purists. For the musician, it is sufficient perhaps to remember that the eighteenth-century oboe would be less penetrating in tone than the modern oboe, and would be somewhat easier to lip, but this advantage would be offset by the acoustical imperfections of the instrument's bore.

The flute has changed, too, in its construction between the early eighteenth century and to-day. Its imperfect intonation was due to the cylindrical bore of early flutes, which does not go well with the series of vibrations set up from an open embouchure. Early transverse flutes (and the whistle-flute had been discarded in concert orchestras at too early a date to need consideration here) with their six finger-holes and a single key, could not be made to play in tune throughout their compass. Not until a conical bore was employed — and later still the acoustically corrected bore of Boehm — could the flute be used in its higher registers without offence to the ears.[1]

So practical considerations had their influence on the use of wood-wind in early orchestral compositions. The wood-wind double the strings, avoiding their high registers, and for a long time violins kept down with them, not venturing beyond the third position. But so different were these two instrumental

[1] Apart from this there was confusion of pitch in Germany, and German players brought their instruments to England.

families in construction that it was inevitable that their distinctions would have to be ultimately exploited. Strings can be played indefinitely without their players getting out of breath. Not so wind instruments, and especially oboes. There had to be periods of rest allowed for the wind, thus leaving the strings with independent phrases that had to be complete in themselves. Whatever, then, may have been the practice among the old shawm-blowing Waits (who had only short tunes to play anyway) it was inevitable that strings should from the first assert their right to be regarded as the foundation of the new orchestra, with the wind doubling them in loud movements and being silent in soft movements. Wood-wind could not, of course, be restricted to such a degrading practice for long: they were capable themselves of a certain degree of loudness and softness, although their scope in this was more limited than that of strings. The next development was to silence the strings for alternate phrases, using strings and wind antiphonally. This, as has already been mentioned, was not a new trick, having been successfully exploited for centuries in vocal music, but in the orchestra it brings into evidence a new and important feature, as it was not for any practical reason, like the breathing of the wood-wind, that the strings were rested, it was done to exploit the charm of a relief from their tone by contrasting it with oboe or flute tone — in a word — colour; and we shall see later how this sense developed still further between the two groups in the concerto grosso — the *concertino* and the *ripieno*.

There remain the brass instruments to be considered. By their nature these were best suited for the larger orchestras and the opera house, although as early as 1692 *The Gentleman's Journal* reported an instance of soft playing on military instruments:

> Whilst the company is at table the hautboys and trumpets play successively. Mr. Showers hath taught the latter of late years to sound with all the softness imaginable.

This 'Mr. Showers' was John Shore, the Sergeant Trumpeter; the man for whom Henry Purcell wrote his florid trumpet obbli-

THE PRE-CLASSICAL ORCHESTRA

gatos. He was succeeded in 1752 by Valentine Snow, for whom Handel wrote trumpet obbligato parts in his operas and oratorios. The king's musicians, in fact, were of some importance in the development of orchestral playing in the seventeenth and eighteenth centuries, for they were in an independent position, with a status to maintain and privileges to uphold. There was a law in force obliging every player on the trumpet, fife, and drum, in any play or show, to hold a licence from the Sergeant Trumpeter, under penalty of 12d. a day on default. Matthias and William Shore, who had preceded John Shore as Sergeant Trumpeters, successfully insisted on their rights of office, and although John Shore and Valentine Snow appear to have been less aggressive in asserting their rights, there can be no doubt that they could be of valuable assistance to musical enterprises by virtue of their office. In 1741 the *London Journal* announced a benefit concert for Valentine Snow, which said: 'At the new theatre in the Haymarket this day will be performed a grand concert of Music by the best hands ... likewise the Dead March in "Saul" to be performed with the sacbuts.' The sacbuts, or trombones, were only to be used in Handel's time by arrangement with the Sergeant Trumpeter, for there were no trombone players in this country except the six in the king's band. Handel wrote for them only three times in his oratorios — in *Saul*, *Samson*, and *Israel in Egypt*.

Florid obbligato parts were characteristic of opera during the transition period between the great days of vocal music and the development of the classical orchestra at the end of the eighteenth century. Their use was to glorify the display of the vocal soloists. In the church music of J. S. Bach and the oratorios of Handel they often became something sublimely beautiful, but they were nevertheless the product of a transitional period, when instruments were treated in a solo vocal style. When the obbligato instrument was a violin, a flute, or an oboe, it sang naturally enough, but when it was a brass instrument the enjoyment of the music could not but be shared with admiration for the player. Had it not been for men like John Shore and

GROWTH OF CLASSICAL ORCHESTRA

Valentine Snow, Purcell and Handel could not have got performances of such difficult trumpet parts — which is equivalent to saying that they would not have written them, for above all else they were practical musicians.

The technique of playing very high trumpet parts, called 'clarino', was established well before Purcell's time. The effect of this endeavour to play florid quasi-vocal parts on the trumpets was to force them up into the top register of the instrument, for below the fourth octave above the instrument's fundamental note a complete diatonic scale was impossible without valves or some other length-adjusting device. Such parts were extremely difficult to play and stood out brilliantly above any other sounds the orchestra was making. They would not blend. But the desire of the new symphonic composers was more often for a full robust blend of orchestral tone than it was for a solo passage, and it was found eventually that the middle register of the trumpet was most suitable for this purpose.

French horns were rough in tone at the beginning of the eighteenth century, but had become established as a most important section of the orchestra by 1750. In 1717 Lady Mary Wortley Montagu said she thought the horns made 'a deafening noise', an opinion that is hard to reconcile with our view of that instrument, founded as it is on the horn effects of later composers. But the horn was but a little removed from the hunting-field when Handel first introduced it at the opera in the Haymarket in 1720. *Radamisto* is the name of the opera in which it was first used. Another, *Giulio Cesare* (1724), is even more deserving of mention, for in it Handel used a quartet of horns to suggest the barbaric character of Ptolemy's Egyptian cohorts. Their use here is very different from the smooth passages for horns such as we are to-day accustomed to hear in works like Weber's *Der Freischutz*.

Handel also recognized the value of horns and clarinets in combination. There are in existence, in the Fitzwilliam Museum at Cambridge, the concertino parts from a concerto

THE PRE-CLASSICAL ORCHESTRA

for two clarinets and *corno di caccia*. It is possible that this work dates from 1724, for as we have seen, two clarinettists seem to have been available for *Tamerlano* in that year, and there were available the horn players used in *Giulio Cesare*. On such circumstances depended the type of composition produced, for Handel, like every other professional musician, had to make the best use of the forces at his disposal: the nineteenth-century habit of engaging a player for an evening in order to play a short passage in one work would have seemed like gross extravagance to the practical musicians of the eighteenth century.

One exception there is to the general rule that the instruments most suited to the new orchestra were of humble origin — the exception is the harpsichord. Even this, or its successor the pianoforte, was doomed to become redundant by 1820, but throughout the eighteenth century the harpsichord performed a dual function in the orchestra. It provided the harmonic body of the music which the melodic instruments clothed and embellished, and it was a centre of control. The harpsichord player, reading from a *basso continuo*, or figured bass part, filled in the essential harmonies, kept up a rhythmic indication of the tempo for the other players, and even filled in a hesitant or missing part. It was a highly-skilled art, this use of continuo, and most of the great composers consequently preferred to direct their performances from the keyboard. Conducting with a baton was apparently *infra dig.* everywhere but in Paris. The composer usually preferred to direct his work from the continuo part, but he was not bound to do so: he could direct it if he liked from the leading first violin desk. In theory every performance was subject to dual control by the principal violin (or leader) and the continuo player, but in practice one of these players would lead the other. A composer performing his own compositions naturally had the most say, whether he was at the harpsichord or the violin.

The picture presented by an eighteenth-century orchestra is, then, of a group of players arranged round a harpsichord, able to see both the harpsichord player and the leading violinist.

The bassoons and 'cellos would be together and the oboes and violins near each other. In many cases they might be called on to play the same parts, or the 'cello, bassoon or double bass, if the orchestra was small, might read the continuo part looking over the shoulder of the keyboard player. The strings would be in numerical superiority over the wood-wind and brass, but not to an extent consistent with our modern ideas of balance of tone. There would be peculiar rhythmic movements of the director's head and shoulders as he sat controlling the rhythm at the harpsichord, and sundry jerks and squirms from the leading violinist trying to do the same thing. In fact, there would be many features about such an orchestra that are familiar to us as we watch a small theatre or café orchestra to-day — the antics of the leader, the scarcity of strings in relation to wind instruments, the habit of playing a missing part on the keyboard or other instrument without any fastidious qualms about its tone-colour; and for much the same reason — they were a practical party of musicians with a job of work to do, having, moreover, to do it with the utmost economy.

All this is understandable, but the likeness may be carried too far. How well we know the mannerisms of the teashop maestro, and his penetrating tone! But we dimly realize that he has to make his tone carry to the extremities of the premises above the noise of teacups and conversation. Such circumstances did not obtain in the eighteenth-century concert-room, though something like them might have applied to the eighteenth-century opera house. When Thomas Mace complained of how 'though Harpsicon, or Organ, or Theorboe-Lute were playing . . . The Scoulding Violins will out-Top Them All', he could have had no idea of the scream of the modern café leader, nor the power of a Stradivarius violin playing the solo part in a modern concerto against a modest background of eighty players, for the violin as Stradivarius made it had neither the range nor volume of that same instrument to-day: the neck has been lengthened to give higher compass, but the pitch has not been dropped

THE PRE-CLASSICAL ORCHESTRA

to accommodate the longer string;[1] consequently the tension has had to be increased, the pressure on the bridge is greater and the tone shriller. To compensate for the increased strain, Stradivarius's original bass-bar has had to be removed and a stronger one substituted. This is the instrument popularly supposed to be exactly the same to-day as when Stradivarius died in 1742.

This striving for greater volume of tone is a somewhat doubtful benefit that has gone on uninterrupted until the present day. Even chamber music has not been free from its influence. Sufficient has been said already to show that the English amateur instrumentalist disliked loud music at the beginning of the eighteenth century; the professional player it was who led the trend towards sonority, aided by the profit-making motive, which led him to think out ways to attract larger audiences. This meant the use of larger concert-halls and larger orchestras, subject, however, to the usual business man's rule of economy, which would lead him to use the smallest number of paid performers necessary for his purpose. It was discovered that a great public demand existed for musical performances employing a large orchestra. The enormous financial success of the final rehearsal of Handel's *Music for the Royal Fireworks* with its orchestra of a hundred players, described on page 28, pointed the way, but the Commemoration of Handel in Westminster Abbey in 1784 set a fashion for monster performances of that master's works which had a considerable influence — for good or for bad — on the course of British music. Such performances depended on the participation of large numbers of unpaid performers, however, and were therefore of more importance in the development of choral music than of orchestral music, but such a division of musical styles was, like everything else in eighteenth-century musical life, less well defined. All the forces used in a mixed vocal and orchestral concert were called 'the orchestra'; to define the instrumental section alone 'the band' sufficed.

[1] It has risen. Since Tartini's day the strain on the strings has increased from 63 lb. to 90 lb.

GROWTH OF CLASSICAL ORCHESTRA

It was not until after the middle of the eighteenth century that the demand for increased sonority began to take control of orchestral performances, and the ultimate effect was all to the good. But it must be admitted that at first the change of style from polyphonic to homophonic was accompanied by a certain laxity in the treatment of the middle parts. 'Formerly', says Henry Purcell, 'they used to Compose from the *Bass*, but Modern Authors Compose from the *Treble* when they make *Counterpoint* or *Basses* to Tunes or Songs.' The melody and its bass were thought out, and given to the first violins and 'cellos (or bass viols) and a part added for the second violins on the treble stave; these would frequently be given also to oboes and bassoons or flutes and bassoons; but what of the tenor part? This was frequently left to be filled in by the harpsichord, or written out for violas in a style that was the very essence of dullness. Lucky indeed was the viola player when the composer did not write a separate 'part' for him, but directed him to double the bass at the octave — at least such a part had interest. The viola, which is the true alto of the string family, had his part taken by the second violins, and in his turn ousted the tenor violin, which is now in consequence obsolete. The tenor violin was held between the knees, like a small 'cello, and was tuned an octave below the violin proper. Such of these instruments as survived into the nineteenth century ended their days as small-sized 'cellos, tuned a fifth below their normal compass and sold to parents who wanted a 'cello 'good enough for the children to learn on'. The exploitation, too, of the A string of the 'cello in its tenor register had in general to wait until the classical symphony was well established.

So much for the instruments themselves, but what of their use as an orchestra? Thomas Mace as early as 1676 had written with all the venom of his age an attack on the new style of composition:

> Our Great Care was to have *All the Parts equally Heard*: . . . But now the *Modes* and *Fashions* have cry'd *These Things* down, and set up a *Great Idol* in their Room; observe with

THE PRE-CLASSICAL ORCHESTRA

what a *Wonderful Swiftness* They now run over their *Brave New Ayres*; and with what *High-Priz'd Noise, viz* 10 *or* 20 *Violins*, &c. as I said before, to a *Same-Single-Soul'd Ayre*; it may be of 2 or 3 *parts*, or some *Coranto, Serabrand*, or *Brawle* (as the *New-Fashion'd-Word* is) and such like *Stuff,* seldom any other; which is rather fit to make a Man's *Ears Glow*, and fill his *Brains full of Frisks*, &c. than to *Season and Sober his Mind*, or *Elevate his Affection to Goodness*.

But Mace was on the losing side. Britain had just emerged from a time when men had shot their neighbours and relations in their wish to establish a society ruled by sober justice with a high reverence for goodness, and a head full of frisks seemed at once an escape from the voice of conscience and a good way of avoiding the consequences of taking morals and politics seriously. Light-hearted entertainment was the vogue, and the orchestra was the latest thing. From Henry Purcell at the Restoration, through the eighteenth century to the coming of Haydn in 1791, musicians supplied that demand and in doing so they grew more proficient in their art, getting to understand the use of orchestral instruments in combination until the end of that century saw the problems of sustained harmony in all parts, and the blending and shifting of tone-colours, brought to their solution in a noble instrument associated with the classical symphony and concerto.

HANDEL'S CONTRIBUTION

HANDEL came to England in 1712, after a prolonged visit to Italy, while still holding the position of Kapellmeister to the Elector of Hanover. He had experienced the normal life of a professional musician on the continent, but preferred to settle in England. Throughout his residence here he was a stranger, German in spirit and Italian in the musical style he tried at first to cultivate in this country. Why then, we may ask, did he voluntarily exile himself from the land of his birth, where he held a secure if modest livelihood? The answer is probably for business reasons: he thought that he could do better for himself as a freelance in this country than as a permanent official in a continental court.

Handel's business was to promote Italian opera, which was then in a precarious state in London, both financially and artistically. Handel knew that he had the skill to produce operas better than those being performed in London, and expected therefore to win fame as a composer and a good reward for his business enterprise. He was not unreasonable in these hopes, but he had to find by experience that grand opera is by its nature unremunerative except to solo singers. On the continent opera was maintained by private patrons; in England patrons were equally necessary, but they were limited in their sense of obligation; in general the opera house was a collective responsibility, and its guarantors would withdraw their support if public favour was not forthcoming; but some men there were to whom a musician could look for support, like the young Lord Burlington, who gathered round them a circle of wits and intellectuals, and the Duke of Chandos had an establishment equal to that of many a continental prince, with his own private chapel and musical staff. Here for three years Handel lived after the collapse of his first efforts to re-establish Italian opera in London. Here he was employed as a private composer

HANDEL'S CONTRIBUTION

similar to any nobleman's Kapellmeister on the continent, but as he was not called upon to produce much purely orchestral music at Canons[1] (the Duke of Chandos' palace) so England missed the opportunity of fostering experiments in the growth of the modern orchestra such as were carried on by Haydn in Austria. But a comparison of Handel's residence at Canons with Haydn's residence at Esterhaz may serve to illustrate the value of a retired and financially secure life to a musician who has experimental work to do in his art.

Haydn, even while he fretted under the necessity of servitude, nevertheless saw the advantages to be had from the control of a musical establishment. In one letter he wrote: 'I am doomed to stay at home. What I lose by so doing you can well imagine. It is indeed sad always to be a slave, but Providence wills it so. I am a poor creature, plagued perpetually by hard work, and with few hours for recreation.' Yet in this busy solitude he could see advantages with which the distractions of city life might have interfered. 'I was cut off from the world. There was no one to confuse or torment me, and I was forced to become original.' So it appeared to Haydn, but such an explanation of his originality is too simple to be entirely satisfactory. Other composers besides Haydn were forced to live retired lives, but few of them showed anything like his originality. The value of Haydn's post with the Esterhazy household was in the resources that lay to hand. 'I not only had the encouragement of constant approval', he wrote, 'but as conductor of an orchestra I could make experiments, observe what produced an effect and what weakened it, and was thus in a position to improve, alter, make additions and omissions, and be as bold as I pleased.'

Handel at Canons was never so isolated as Haydn at Esterhaz, for he was within easy distance of London. He could call in at Thomas Britton's loft to play there his latest harpsichord pieces, and he could find time and opportunity to accompany

---

[1] His Overture in D minor, however, which Elgar has arranged for modern orchestra, belongs to this period.

the royal barge up the Thames, directing the performance of the numerous dances and instrumental airs that are known to us as the *Water Music*. Indeed, Handel never seems to have lost an opportunity of courting favour with influential Londoners, which indicates that he had always in mind a return to the public life of a London theatrical manager; but while he was at Canons his duties were similar to Haydn's at the Esterhazy establishment: he had to compose and perform music for his employer's gratification, and in so doing was able to make experiments.

The music required of Handel by the Duke of Chandos was mainly choral. The Duke's private chapel, although flamboyantly Italian in its decorations, was nevertheless a temple of Anglican Protestantism. For this establishment Handel wrote the Chandos Anthems — grandiose works in keeping with the general atmosphere of Canons — but they led Handel to a new style of composition, different from the German Lutheran style of his *Brockes' Passion* and different also from the Italian style of his operas. The Chandos style led by way of *Esther* — first produced as a masque at Canons and later revised as a secular oratorio — to the grand choral works which appealed so strongly to the spirit of English Protestantism and formed the basis of English choral taste during the nineteenth century. Handel had opportunities for experiments at Canons equal to, but different from, those of Haydn at Esterhaz, and it is significant that his stay there resulted in a development of technical resources which, like Haydn's, were in their final stages somewhat wasted on the requirements of a single patron, but stood him in good stead when the necessity of composing for a large and critical public audience arose.

It is equally true that even in his 'secure' years, when the bounty of some private patron like Lord Burlington or the Duke of Chandos relieved Handel of the financial anxiety attendant on catering for a fickle public, his best work was produced for his most critical audiences, and those whose favours had to be won. The *Water Music* is a product of his early years

HANDEL'S CONTRIBUTION

in London, yet it is among the most pleasing of his music for the orchestra. The circumstances of its production are well known; it was written to serenade a wealthy patron — in this case King George I. Such serenades were a recognized way of attracting the attention of a possible patron, and most eighteenth-century composers had occasion to write them. They could be played by a small group of musicians outside the house of the prospective patron, and, because they had to be played in the open air, these eighteenth-century serenades employed, as a rule, more wind instruments than a composition intended for an indoor orchestra. The movements varied in number, and were generally in dance-forms with one or more songlike slow movements among them, while the first and last movements were usually in march time — an indication that they might have been played as the band approached and retired from the house. But not all serenades of this type were written to attract the attention of a prospective patron; some were commissioned. In London the ideal place for such serenades was the river, a busy thoroughfare in the eighteenth century, far more pleasant than the streets, and quieter. Here it was usual for a wealthy pleasure-party to hire a second barge carrying a band of musicians whose duty it was to play lively music, keeping near enough to the luxurious barge of the guests for the music to be heard. In Mrs. Delaney's *Correspondence* such a journey is described: 'We rowed up the river as far as Richmond, and were entertained all the time with very good musick in another barge. The concert was composed of three hautboys, two bassoons, flute allemagne, and young Grenoc's trumpet.' George I was very fond of the river, and the story goes that the Baron Kielmansegg (Master of the King's Horse and husband of the king's favourite but forbidding mistress, whom Handel had previously met in Venice) and the young Lord Burlington persuaded Handel to write a series of gay dance movements, and themselves hired the musicians to be rowed behind the king's barge playing this music. The story is plausible, but the date given by Hawkins on the authority of a friend of Handel's cannot be confirmed from any other source, and Handel's manuscript of the *Water Music* cannot be found. Hawkins gives the date as 1715, but the earliest report of Handel's *Water Music* appeared in 1717 in the *Daily Courant* of July 19th:

> On Wednesday evening at about eight the King took water at Whitehall in an open barge, wherein were also the Duchess of Bolton, the Duchess of Newcastle, the Countess of Godolphin, Madam Kielmanseck, and the Earl of Orkney, and went up the river towards Chelsea. Many other barges with persons of quality attended, and so great a number of boats, that the whole river in a manner was covered. A City Company's barge was employed for the music, wherein were fifty instruments of all sorts, who played all the way from Lambeth, while the barges drove with the tide without rowing as far as Chelsea, the finest symphonies, composed express for this occasion by Mr. Hendel, which His Majesty liked so well that he caused it to be played over three times in going and returning. At eleven His Majesty went ashore at Chelsea, where a supper was prepared, and then there was another very fine consort of music, which lasted till two, after which His Majesty came again into his barge and returned the same way, the music continuing to play until he landed.

The suggestion first made by Streatfield in 1909, that Handel's *Water Music* was not written for any one river trip, but is a collection of pieces used for various river trips, has been accepted by the editor of the new edition of this music published (1943) by Boosey and Hawkes, who, relying on the edition in parts issued by Walsh in 1732 or 1733, points out that the music falls into three groups. First a group of nine pieces in F major and the relative D minor, scored for 2 oboes, 1 bassoon, 2 horns in F, and strings; then a second group of five pieces in D major scored for 2 trumpets, 2 horns, 2 oboes, 1 bassoon and strings; finally a third group of six pieces in G major and G minor for strings with additional flutes. His is a theory based on practical information, and may well be true, but there must always be room for doubt where eighteenth-century publishers are the

HANDEL'S CONTRIBUTION

source of authority and not the composer's manuscript. Chrysander, in Vol. 47 of his Handel-Gesellschaft edition, has short shrift for Walsh, for he says:

> Walsh printed only nine of the twenty pieces of which this work consists, in ten divisions ... What his seven instruments give is also incomplete with regard to the number of instruments. The small value of this publication is still further diminished when the musical contents of the several parts are tested. The violins and hautboys are combined in the same part, but when they part company the hautboys always have too little given them. The trumpets are not mentioned by Walsh, but some of the notes belonging to them are put into the horn parts; it is unintelligible how any rational mode of playing was possible under these circumstances. The two horn parts are given by him in C major; but the fifth movement ... has over the first horn part the prescription in German 'D Horn' and over the second in German and English 'D Horns'. This publication of Walsh is perhaps the least reliable of all the instrumental works which that energetic but unconscientious publisher put forth in parts. Thus the original parts of Handel's score are not to be gathered from it.

Chrysander's view, however, is too reminiscent of the nineteenth century: one has to be lenient with eighteenth-century publishers, even as with composers, who set down as little as possible on paper. Walsh, of course, was careless. He describes the work on his title-page as 'The Celebrated Water Music in Seven Parts, viz. Two French Horns, Two Violins or Hoboys, a Tenor and a Thorough Bass for the Harpsichord or Bass Violin composed by Mr. Handel'. A simple sum of addition will reveal that there are here only six parts — Walsh has not mentioned the bassoon on his title-page, for which a part is included. But for practical purposes it does not matter — the bassoon doubles the bass. The oboes and violins reading off similar parts were not doing anything unusual for the early eighteenth century, and trumpet parts cued into horn parts save both paper and labour. Such was the eighteenth-century view, and players and directors of music understood.

The practical view that the *Water Music* contained in Walsh's first edition in parts represents the music given on three different occasions has much to recommend it. The key of F was most suitable for good horn tone and the key of D for trumpets. The first group in F may possibly have been used at a river party at which trumpets were not to be used; the second group in D may have been used at the 1717 river party reported in the *Daily Courant*, for there is an additional report by Bonet, only discovered in 1922, confirming the contemporary report that there were fifty players, and mentioning also that they comprised 'des trompettes, des cors de chasse, des haut bois, des bassons, des flutes allemandes, des flutes françaises a bec, des violons et des basses, mais sans voise'. It is interesting to note that by 1783, when Dr. Samuel Arnold commenced his edition of Handel's music, the *flutes françaises a bec* had become *flauti piccoli*.

The scoring of the third group of pieces, i.e. strings and flutes, does not suggest that they were composed expressly for outdoor use. They may, of course, have been played during the supper at Chelsea, but such surmise is not really called for. It assumes too readily that because pieces were published under the title of *Handel's Water Music* that they must therefore have been composed expressly for use on the river. This is awarding the eighteenth-century publisher more marks for integrity than he is likely to have earned; we shall do well to remember that the *Water Music* was Handel's most popular instrumental success in his lifetime, and that would be enough inducement for any publisher of that time to sneak in a few odd pieces under the same title.

All the movements are gay, even the slow airs having happy moods. They are in the simplest of forms, the dances keeping within the scope of their original purpose, without any pretensions to cleverness. Some of the movements, however, are directed to be played thrice through, which means that oppor-

HANDEL'S CONTRIBUTION

tunities could be taken for contrasting tone-colours. One of the minuets in a manuscript by Handel's amanuensis, J. C. Smith, is directed to be played three times — first by trumpets and violins, then by oboes and horns, and finally by the full orchestra. It does not necessarily mean that because a movement is written out in all parts that all parts were obliged to play it at every repetition. Parts were often copied in a hurry, and in any case had to be copied or printed as cheaply as possible, and the direction 'three times' written on the page served its purpose. The composer was present to decide which instruments were to be used, and even if the composer was not there, the players were capable musicians, easily able to make up their own minds on such a matter.

The orchestra for the *Music for the Royal Fireworks* was one hundred strong, made up of 42 strings against 56 wood-wind and brass. There were actually 24 oboes, 12 bassoons, 1 serpent, 9 trumpets, 9 horns, and 3 drums. According to our modern ideas therefore the strings would be swamped by a mass of wind tone, most of which was liable to be out of tune. The eight movements were in the key of D (again probably on account of the trumpets, for the natural trumpet in D gave the best all-round results). The movements are: Overture, Allegro, Lentemente, Bourree, Largo alla Siciliana, Allegro, and two Minuets; they had the advantage of the widest publicity, but never gained the popularity enjoyed by the *Water Music*. Their final rehearsal at Vauxhall Gardens, a week before the actual firework display for which they were written, had a far greater triumph than the actual performance, when the fireworks were only partly successful as such, and the central pavilion — which was not intended to be burnt — caught fire and blazed merrily. The short pieces that comprise Handel's music for this event were intended to be played between the 'set pieces' of pyrotechnical art; but with the general failure of the intended effects, and the excitement of the unintended (which culminated in the arrest of Servandoni, the architect, who drew his sword on the Controller of the Ordnance when he saw his

GROWTH OF CLASSICAL ORCHESTRA

ornate pavilion in flames), it is hardly a source of wonder that Horace Walpole, in describing the scene, fails to mention Handel's music. It is highly probable that in spite of the enormous size of the orchestra, few of the spectators heard much of it. By the nature of the occasion there was a gap (by no means silent) between each movement of Handel's suite; modern arrangements of this music as a concert work are obliged to treat it as a complete whole, instead of a series of pieces composed to illustrate various displays of fireworks, and the difficulty of forming these pieces into an effective *concerto grosso* is much the same as in making an orchestral suite out of the incidental music to a play.

The *Water Music*, on the other hand, is more closely akin to the eighteenth-century idea of a suite, and the modern edition issued by Boosey and Hawkes in 1943 describes the first nine movements of this music as a *concerto grosso*. It is not inconsistent with that form, although Handel, had he been writing such a work for concert performance, would probably have reduced the number of movements. *Concerto grosso* form was fluid, and the number of movements and constitution of the orchestra varied. In general this was because the composition was written to suit the instruments available, and not, as the modern habit is, to engage players in accordance with the demands of the score. The composer had not, in the eighteenth century, receded into a remote hinterland from which his ideas could come to the conductor only through the medium of certain marks on paper. Certain successful works could be, and were often, repeated, but a study of the lives of eighteenth-century composers makes it evident that their great concern was not how to get repeated performances of their works by other musicians so much as to keep up the supply of music for their own immediate needs.

Handel was oppressed by this necessity all his life, and frequently resorted to a rearrangement of some earlier work to satisfy an immediate demand for something new. Sometimes a movement appears several times in different works, as, for

HANDEL'S CONTRIBUTION

example, the third Organ Concerto, where the first two movements are the same as the corresponding movements in the fifth Sonata for strings, and the finale is identical with the finale of the second Sonata for Flute, and has its origin in a song for soprano voice with 'cello obbligato *Non ho cor che per amatti* from the opera *Agrippina*. The fifth Organ Concerto in F is an arrangement of his eleventh Sonata for Flute, throughout, and the sixth Organ Concerto is similarly an arrangement of one composed for the harp. These are accordingly slighter works than those — like the first Organ Concerto — conceived originally for the organ. In the sixth Organ Concerto the accompanying strings are muted and the basses marked 'pizzicato' so as not to overpower the harp in the original version, while the *Siciliana* of the fifth Organ Concerto remains a reminder of its original conception for the flute.

The term Organ Concerto is here used because that is the name under which modern arrangements are usually issued, but Handel's publisher, Walsh, called them 'Six Concertos for Harpsichord or Organ, Op. 4' and 'A Third Set of Six Concertos for the Harpsichord or Organ, Op. 7'. The latter set Walsh issued after Handel's death. It is to be hoped that more attention will be paid to the alternative instrument in the near future, for some of the first set are far better fitted for a harpsichord or pianoforte than for an organ. It is unfair criticism to remark on the inferiority of Handel's Organ Concertos in comparison with J. S. Bach's organ works, when in some cases Handel was not writing expressly for the organ; and even when he was, the organ he used had no pedals. Criticism must be made, not in accordance with modern ideas, but with those of Handel's time, which were directed to practical issues. Only one organ in London had pedals — that was in St. Paul's, and possibly Handel's seventh Organ Concerto, which requires an organ with pedals, was written for this organ, although opinion generally is that Handel wrote it for use during one of his visits to the continent. Handel's use of the pedals in the climax of an *Andante* on a ground bass in this concerto is evidence of his practical musicianship and of his instinct — common enough in his time — to make the most artistic use of any resources at his disposal. So it must have been with his sixth 'Oboe' Concerto, which includes an obbligato part for harpsichord: the finale of this came in useful several times, for it is found also in the third *Suite de Pieces*, in the overture to *Pastor Fido* and as the finale of the tenth Organ Concerto.

The six so-called Oboe Concertos need not detain us long, since they belong to Handel's pre-English period,[1] but they were his first efforts in the style of the *concerto grosso*, and are closer to the general conception of that form than his later concertos. The term 'Oboe' is a misnomer, for although there are examples of the use of a solo oboe to be found in the *Largo* of the second of these concertos and in the *Andante* of the fourth, the oboes do not predominate in other movements of these works to the extent the title might suggest. The reliance placed on double-reed instruments by composers for the early orchestras has already been mentioned — Handel's 'Oboe' Concertos are scored for the usual strings with 2 flutes, 2 oboes, and 2 bassoons: the commonest orchestral combination, in fact, of that time. The misleading factor in this description lies in our modern bias towards thinking of such an orchestra in terms that imply familiarity with the orchestra of Haydn and Mozart, who used the wood-wind either as soloists or as a choir, according to choice. The use of instruments in a *concerto grosso* was quite different. In his first 'Oboe' Concerto Handel detaches two oboes and a violin from the main body to sustain individual parts; in the second movement one of the oboes drops out and two flutes are added, so that there is now employed, against the general background of strings, a group of four soloists, comprising one oboe, one violin and two flutes: in the third movement a return is made to the disposition of the first movement. In modern parlance, the style is a combination of chamber music and orchestral music: the small group of soloists was called the

[1] They were composed for the Prince of Orange and Princess Royal, and contain material used in lessons.

HANDEL'S CONTRIBUTION

concertino and the accompanying body the *ripieno*. The whole was a *concerto grosso*. There was no reason why a single instrument should not form the *concertino*; such a practice need not change the character of the work so much that it would cease to be a recognizable *concerto grosso*, but this was not usual in the latter half of the seventeenth century and early eighteenth century, when the 'Oboe' Concertos were written: it grew in favour, however, as the eighteenth century proceeded, and Handel anticipated future developments in his drift towards a harpsichord or organ *concertino*, to which subject we shall have consequently to return, but before doing so mention must be made of Handel's twelve Grand Concertos in which the combination of a small group of soloists and a large *ripieno* reach their highest level in this master.

The prolonged neglect to which the Grand Concertos have in the past been subjected in England is partly to be explained by the peculiar reputation that Handel acquired soon after his death; a reputation for hearty religious music. His works have been. applauded by moralists and very widely spread by organists and choirmasters. For this reason the Organ Concertos got to be known when the Grand Concertos for strings were hardly known at all. It is the task of the orchestral conductor to make known the Grand Concertos; yet here again there is a difficulty, for the pre-classical concerto needs a different approach from its successors, and for too long the general policy has been to arrange pre-classical works for a post-classical orchestra. In the course of our development of orchestral sense, however, the British Handelian has been too often a powerful reactionary, and probably for that reason orchestrally-minded people have paid more attention to the concertos of J. S. Bach than to those of Handel. Fortunately they are now being revived. Written entirely for strings, they have a scope of expression that would give the lie to anyone artless enough to think that all-string tone lacks variety. Against a strong *ripieno* Handel presents in each of these works a *concertino* of two violins and one 'cello. With such forces Handel's range of effect is truly remarkable: the melancholy of the sixth Grand Concerto contrasts well with the spontaneous gaiety of the seventh. By an amazing variety of devices Handel adds to the interest of his themes. Consider, for example, the key sequence of the ninth Concerto in F. After a dignified *Largo* in three-four time, the tempo changes to a brisk *Allegro*, in four-four time. Then the key changes to D minor for a dainty *Siciliana, Larghetto*, in six-eight time, and back to F major for a bold and vigorous fugue, *Allegro*, in four-four time. The next key change is to the tonic minor (F minor) for a graceful minuet, and swings back again to F major for the last movement, a rollicking jig in twelve-eight time. Handel strives for variety, too, in the sequence of moods in his several movements, which are more numerous in the Grand Concertos than in the 'Oboe' Concerto, but less numerous than in the *Water Music* and *Fireworks Music*, for the obvious reason that the two last were written for an entertainment spreading over a long period of time — an afternoon or an evening, whereas the concertos were written to be played through immediately, as a related sequence of movements.

Much material from the Grand Concertos was used for Handel's second series of six Organ Concertos, which have not for that reason been commented on here, but number two of this series deserves mention because of its introduction of the Cuckoo and Nightingale, much in the same way that Beethoven introduced these bird-calls into his *Pastoral Symphony*. They do not appear in Handel's original Grand Concerto version, but only in the Organ version. The reason is probably to be found in Handel's intention for their use. The Grand Concertos were dignified works intended to form the strong meat of the concerts for which they were respectively written, but the Organ Concertos were played between the acts of his oratorios as a relief from the tension of the latter. There is, then, good reason for the introduction of ear-tickling novelties in the Organ Concertos which would be out of place in more serious works.

It is probably because of the association of the **Organ Concertos** with the oratorios that these concertos continued to be used

HANDEL'S CONTRIBUTION

when the others were neglected. There are few indications of the particular oratorios with which the various organ concertos were associated—probably they were not associated with any in particular, but it is interesting to note that the last movement of the second Organ Concerto is a development of the organ obbligato from the song 'In the Battle Fame Pursuing', in *Deborah*, and that the graceful minuet which terminates the Organ Concerto in B Flat, No. 3 of the third set, is known as the *Minuet from Esther*, from which it may be inferred that this was once, at least in the popular mind, associated with that oratorio. In the main, however, these surmises count for little beside the importance of Handel's experiments in the use of the cadenza, later to become so important a feature of the classical concerto, and his one experiment of a choral ending to an instrumental work, by which he anticipated Beethoven.

There are no cadenzas in the first six of Handel's Organ (or Harpsichord) Concertos, Op. 4, published in 1738, but in a third set of six published as Op. 7, posthumously, every concerto except the fifth has one or more breaks in the score, filled in by the words *organo ad libitum*. The first of these concertos, in B flat, is the magnificent work previously described as having a ground bass wherein Handel employed pedals for the only time in his life. That is in the second movement, *Adagio*. The first movement is a brilliant piece in common time which breaks off suddenly at its climax with the direction *organo ad libitum* in the blank space on the score: following this come eight bars to bring the movement to its close.

The second concerto of this set has two such breaks in the main movement; one about three-quarters of the way through, and the other at the end. In the third concerto Handel brings his first movement to a full close, and then gives his direction for organ to play extempore, and in No. 4 in D minor he uses this device repeatedly, for in the principal movement he silences the orchestral parts no less than six times to allow for extemporization on the organ. In the sixth concerto of this set, all the previously-tried devices for introducing a cadenza are

GROWTH OF CLASSICAL ORCHESTRA

employed: there is an *organo ad libitum* space between the first and second movements, and in the second movement itself *a tempo ordinario* in B flat, duple time, there are three *ad lib.* spaces, each led into by a solo bar from the organ. The orchestra would no doubt pick up the thread of their parts at a sign from the composer.

The fifth concerto has an *organo ad libitum* space in the fourth movement, after four bars of introduction, and for its second movement, an *Adagio* consisting of three bars of figured bass. There is another effective set of variations on a ground bass in this concerto (besides that mentioned in No. 1 of the Op. 7 set), the fourth movement, marked *Andante Larghetto e Staccato*.

The explanation for these extempore interpolations is partly artistic and partly utilitarian. Handel was an able improvisor, and this of itself would be sufficient justification for his development of this technique in his concertos, but in addition it must be remembered that he went blind in his last years, and had therefore to rely more on his skill in improvisation than he might otherwise have done.

Handel's anticipation of Beethoven's choral ending to an instrumental work is to be found in an autograph copy of the *Fourth Organ Concerto* in the British Museum, which has an Alleluia chorus founded on a theme in the concerto. It was written in 1735 and used as a conclusion to the oratorio, *Il Trionfo del Tempo e della Verità*, an early work revised in 1757.

Apart from these experiments, there are some most interesting features in Handel's orchestration in the organ concertos. The opening of the fourth concerto of the third set, with a lovely singing passage in D minor for 'cellos and bassoons well up in their tenor register (Handel here uses the tenor clef) anticipates nineteenth-century practice. The simple but effective colouring of the Gavotte in G minor in the fifth concerto of the third set is also noteworthy as typical of the best contemporary practice. After fourteen bars for oboes and bassoons comes the direction 'Viol. sen. Hautb.', and 'Violone e Viola sen. Bassoons'. From bar twenty-seven onwards the Gavotte is played *tutti*. This simple

HANDEL'S CONTRIBUTION

manipulation of tone was effective according to the limitations of the period, and if it leaves something to be desired to modern ears, let us not forget that the primitive orchestral effects of Peri and Monteverdi would have sounded equally crude to Handel, but he nevertheless sought to employ their principles of characterization by tone-colour more effectively in the light of his later knowledge of the possibilities of orchestral instruments. His use of obbligato is one of the chief joys of his oratorios, and in all his stage works Handel's emotional range is greater than some conductors have in the past realized. Even in his Italian operas, where he was tied to formal vocal styles in a manner wasteful to one of his genius, he makes use of the orchestra to infuse emotional interest into the stage situation. *Giulio Cesare* has a pallid plot into which Handel tried his best to instil life by his music. It is probably the first opera to employ four horns in the orchestra, and the passage where Cleopatra shows Caesar a vision of Parnassus is accompanied by harps, viola da gamba and theorbo. The principle is that of Monteverdi — to create a distinctive emotional atmosphere by a distinctive tone-colour. The storm-music in *Riccardo* carries conviction, and there are plenty of instances of associated tone-colourings in this opera: a 'bird-song' for soprano has a florid piccolo accompaniment, while a bass flute[1] is used to accompany a beautiful if mournful air when a despairing lover prays for death. Trombones (or sackbuts) he used in *Samson* and *Saul*; but the scarcity of players on that instrument probably prevented him from making more general use of them. *Saul*, however, has an independent part for an organ — a very different thing from Handel's general practice of playing the *continuo* on the organ instead of on the harpsichord. A chime of bells accompanies the rejoicings of the Children of Israel, David exorcises the evil spirit to the accompaniment of a theorbo, and Samuel's ghost speaks to the accompaniment of two bassoons.

The Messiah affords most opportunities in these days of

[1] The Bass Flute is really an alto instrument.

GROWTH OF CLASSICAL ORCHESTRA

studying Handel's oratorio style. Unfortunately the orchestration has been so altered by additional accompaniments that it is difficult to guess what went on in Handel's orchestra. There is a restraint about *The Messiah* that gives the work a special devotional appeal much more fitted to this subject than Handel's usual operatic style. He avoids the theatrical and the obvious in his orchestration. The *Pastoral Symphony* uses strings, not oboes, cornetts, chalumeaux, or clarinets, with which he clothed other pastoral music. It is marked 'pifa' in the original, which means that it was after the style of an early Italian bagpipe tune. Mozart introduced wood-wind to this music to make it sound like the bagpipes, surely an error of taste, for it is the simplicity of the shepherds that Handel wished to emphasize, not their instrumental prowess. The use of the word 'symphony' here is a continuation into modern times of the eighteenth-century use of the word.

Most important of all is Handel's use of obbligato. The dramatic trumpet obbligato to *The Trumpet Shall Sound*, set against a stormy background of strings, is an arresting example, but the restrained beauty of the obbligato to *How Beautiful are the Feet* or *Rejoice Greatly*, both given to the violins in unison, show him to be a musician of fine taste as well as technique. His only peer in the use of this device was J. S. Bach.

Handel died in 1759, but the Handelian tradition, fostered by Boyce at the Three Choirs Festivals, and by Burney after the Commemoration of 1784, was the beginning of a new faith in musical Britain.

THE ENGLISH MANNER

HANDEL towers above his English contemporaries like a Gulliver among the Lilliputians, but the smaller men, as in Swift's tale, were sincere and earnest in their own work, however light and airy it might be. Burney is responsible for much neglect of Handel's instrumental contemporaries, yet he is not to be blamed, for he was too near the scene to view the whole in true proportion. Later writers copied Burney's opinions with too little consideration for his prejudices, which are obvious enough in all conscience, and when at last the revolt against Burney came, with our appreciation of English Tudor music, the pendulum of criticism swung too far in the opposite direction, and poor Burney has been libelled by otherwise careful critics. Yet Burney and Hawkins are invaluable as contemporary recorders of eighteenth-century music, and contemporary prejudices are as valuable as contemporary records.

There was something in the lesser men that Handel could not improve upon, and that was their gift for graceful airs. Arne, Boyce, and Greene have not the profundity of Handel, but they have a melodic charm that is closer to the English countryside than the cosmopolitan Handel was ever able to get. How the picture has been distorted can be seen from the following criticism of the character of Maurice Greene, taken from George Hogarth's *Musical History, Biography, and Criticism*, published in 1835:

> When Handel first arrived in England Greene courted his society with great assiduity; but a violent animosity afterwards took place between them, which lasted during Greene's life. In consequence of this, Greene took every opportunity of decrying the works of Handel, and extolling those of his rivals, while Handel always expressed aversion and contempt for Greene. 'Handel', says Burney, 'was too prone to treat inferior artists with contempt. What

provocation he had received from Greene after their first acquaintance, I know not; but for many years he never spoke of him without some injurious epithet.'

The 'provocation' was that Greene had courted Buononcini's friendship as well as Handel's. Greene's friendship for Buononcini, in fact, got him into more serious trouble with the Academy of Antient Music, for he was instrumental in procuring there a performance of a madrigal purporting to be by Buononcini, but strongly suspected to be by Lotti. This was the reason for Greene's secession from the Ancient Academy and founding of the rival club at the Devil Tavern (called the Philharmonic Society), which is said to have brought the remark from Handel, 'Toctor Greene has gone to the Devil'.

Greene, in fact, tried to be friendly with everyone in his profession and was especially helpful to Boyce.

The association of these minor English composers indicates a sympathy for each other's work and a desire to help their less fortunate brethren, which does not bear out Hogarth's statement, for both Greene and Handel worked hard in the cause of the Royal Society of Musicians, a society established by Festing, the violinist (with whom Greene was associated in forming the Philharmonic Society at the Devil Tavern). One day as Festing and his friend Weideman, a flautist, were walking through the filthy streets of London, they saw two small boys driving milch asses. Struck with their poor appearance, they were even more concerned when they discovered by questioning them that the boys were the orphaned sons of an oboe-player named Kytch. They at once organized a subscription in their profession to relieve these children from their distress, and set about forming a permanent society to care for musicians and their families who had met with misfortune. 'There was', says Burney, 'no lucrative employment belonging to this Society, excepting small salaries to the secretary and collector, so that the whole produce of benefits and subscriptions is net, and clear of all deductions and drawbacks.'

THE ENGLISH MANNER

Among its first members were, besides Festing and Weideman, Handel, Boyce, Arne, Christopher Smith, Carey, Cooke, Edward Purcell, Leveridge, Greene, Reading, Hayes, Pepusch, and Travers. That was in 1738. By the end of the next year the Society had 226 members, including all the most eminent musicians of the time. Handel took a special interest in this Society; he gave concerts and composed concertos for its benefit, and on his death bequeathed to it the sum of £1000. The Royal Society of Musicians was, in fact, one of the most worthy charities inaugurated in the eighteenth century.

There was need of such an organization, for orchestral players had a hand-to-mouth existence, with no more assurance of regular employment than an artisan engaged in a seasonal occupation. They were anonymous hirelings employed by composers and managers trying to make a profit from a fickle public. When, for example, in 1733 the Vice-Chancellor of the University of Oxford invited Handel to perform *Athalia* in connection with the University Commemoration festivity, he met with opposition. 'It was intended to have the drama as well as music represented at the Festival, but the Vice-Chancellor would not permit the actors to come to Oxford, and not without justice, although they might as well have been here as Handel with his beggarly crew of foreign fiddlers.' The quotation is taken from Hogarth, who bowdlerized the passage, for Thomas Hearne actually wrote: 'Handel with his lousy crew, a great number of foreign fiddlers.' Hearne was the most prejudiced of men. Quite unmusical, and a hard-bitten Jacobite who always referred to the King as the 'Duke of Brunswick' (and Handel was firmly established in Hanoverian favour). 'Lousy' might be an exaggeration, 'beggarly' is highly probable, for Hearne and the undergraduates objected to paying five shillings for seats at the performance, though this was only half the price usual in London; but in the end most of them paid their five shillings and went to 'try how a little fiddling would sit upon them', for it is on record that 'notwithstanding the bar-

barous and inhuman combination of such a parcel of unconscionable scamps, he (Handel) disposed of most of his tickets'. And it was true that Handel, like many other German musicians of his day, was careful over money payments.[1]

It would appear from a letter by William Harris to his sister, written in 1746, however, that Oxford opinion on the status of Handel's orchestral players was not groundless, for Harris writes: 'Yesterday morning I was at Handel's house to hear the rehearsal of his new *Occasional Oratorio*. It is extremely worthy of him, which you will allow to be saying all one can in praise of it. He has but three voices for his songs — Francesina, Reiholt, and Beard; his band of music is not very extraordinary. Du Feche [i.e. William Defesch] is his first fiddle, and for the rest I really could not find out who they were, and I doubt his failure will be in this article.'

Anonymity in performers is a bad sign. Had any of these players been known to the music-going public Handel would have made the most of whatever good reputation they had earned, for he was an astute showman. They must have been players who had failed individually to make their marks on the public, which means that they were dependent on whatever hack-work they could obtain under the management of speculative producers, on whatever terms they could get. Only by establishing a distinctive reputation with the public could a musician obtain any power of bargaining with his employer. The wages of these anonymous orchestral players would be low, but even so, they were better off than continental players, of whom London attracted many from their native countries, for Leopold Mozart, writing in 1764, remarks on the extraordinary expense of an English orchestra.[2] Teutonic parsimony may have had something to do with this opinion, however, for he made a good profit on the concert, and qualifies his criticism with the statement that 'most of the musicians would take nothing'.

[1] When the University offered him a doctorate he refused with a characteristic phrase: 'Vat te tevil I trow my money avay for dat vich de blockhead vish?'
[2] The usual fee for string players in the latter half of the century was half a guinea for a single concert.

VIEW OF THE ORCHESTRA AND PERFORMERS IN WESTMINSTER ABBEY AT THE COMMEMORATION OF HANDEL, 1784

A CONCERT TICKET BY HOGARTH

THE ENGLISH MANNER

This willingness of musicians to help each other is one of the most gratifying features of their history. So general had it become, even in the eighteenth century, that a tradition had been established which persisted right up to modern times. A player in need would be assisted by the proceeds of a benefit concert, often a superior affair, at which all who took part did so without a fee. Similarly, any charitable object could be assisted by the proceeds of a concert, but such concerts were so numerous that players could not regularly waive their fees, especially as charity concerts were more fashionable and expensive than any others. The advertisement value of charity, however, was high; to quote again from Leopold Mozart, when he allowed his children to appear at a charity concert at Ranelagh in 1764: 'I have permitted Wolfgangerl to play the British patriot and perform an organ concerto on this occasion. Observe, this is the way to gain the love of the English.'

Various charities relied as much on a musical attraction as upon the worthiness of their cause, and some even came to hold annual meetings at which the same music formed a regular feature. One of these was the Corporation of the Sons of the Clergy, founded as early as 1655 to relieve distress among clergymen and their families and educate their children. Each year a festival is held (for the Corporation still exists), consisting of a choral service with a sermon, followed by a dinner. Certain compositions have been at various times associated with these meetings, and orchestral accompaniment became general in 1698.

Purcell's *Te Deum* and *Jubilate in D* were given until 1713, when the influence of Handel on English taste began to show itself in these services with the adoption of that composer's *Te Deum* and *Jubilate* (composed for the Peace of Utrecht), taking turns with Purcell's compositions in alternate years. This policy continued until 1743, when all these four compositions were displaced by Handel's *Dettingen Te Deum*. From 1720 throughout the eighteenth century Handel's overture to *Esther*

GROWTH OF CLASSICAL ORCHESTRA

was used as a prelude to these services. The annual festivals of the Corporation of the Sons of the Clergy, in fact, indicate as surely as anything the trend of popular taste in the eighteenth century, which, strongly in favour of Henry Purcell at the beginning of the century, was influenced by Handel as soon as he came to this country, and by 1740 was completely given over to that master's influence, except for some temporary attention to Boyce, who for a time conducted the music at these meetings.

It must be allowed, however, that charity concerts and sermons did little to advance the cause of music. Their object was to raise funds for other purposes, and their policy was to repeat almost *ad nauseam* the music that had proved on previous occasions to be attractive. The Three Choirs Festival was an eighteenth-century offshoot of the annual festival of the Corporation of the Sons of the Clergy. Dr. Bisse, Bishop of Hereford, had preached at one of the Sons of the Clergy meetings at St. Paul's Cathedral. Like the London festivals from which they sprang, the Three Choirs Festivals rendered Purcell's *Te Deum*, alternating annually with Handel's *Te Deum* for the Peace of Utrecht. In due course these were superseded by Handel's *Dettingen Te Deum* exactly as they were in London. The fund raised was devoted to the same objects as the London Festival. Important as the Three Choirs Festival has become, it cannot be claimed that it did anything more than follow London tastes in the eighteenth century. They even went to the extreme of appointing the same conductor as the London festival and allowing him to introduce London choristers to the Severn Valley performances.

William Boyce was the conductor named. He commenced his duties at the Three Choirs Festivals in 1737. Boyce had been able to introduce some of his own compositions to the public through the London festivals of the Sons of the Clergy, and was not long in introducing further compositions of his own and his friend Greene to the Severn Valley festivals. The festival was of two, and later, of three days' duration,

THE ENGLISH MANNER

and permitted secular concerts as well as the cathedral service. At the secular concerts, Boyce conducted by beating time with a roll of paper, instead of following the usual practice of directing the playing from the harpsichord. Soon Boyce and Greene's compositions began to alternate with Handel's at the Three Choirs Festivals. Greene's pastoral *Love's Revenge* and Boyce's pastoral *The Shepherd's Lottery* were given, sandwiched between many layers of Handel works, including *Acis and Galatea, Samson, Judas Maccabaeus,* and *The Messiah.*

Such organizations used music as a means to an end, but music itself was not that end. The trend of charity concert programmes indicates the trend of popular taste among the upper and middle classes, and this was away from highly-elaborate vocalism towards music with a meaning. This meaning was best discerned in the sturdy oratorios of Handel, and found its most popular expression at the Handel Commemoration of 1784. London led the provinces in all matters of taste: the annual gatherings of the Corporation of the Sons of the Clergy were reflected not only in the Three Choirs Festivals but in the Birmingham Festivals which started in 1768. William Boyce had the task of directing this music both in London and in the Severn Valley, but it is possible that he was out of sympathy with the trend of taste towards massive effects, for his best work is in slighter forms, designed to entertain rather than to elevate, and found its most happy expression in his eight symphonies.

Boyce's symphonies cannot claim to have the power of Handel's Grand Concertos nor even of his Organ Concertos. Intellectually they are about on a par with Handel's so-called 'Oboe' Concertos. They have, however, a melodic charm that stands in the English tradition. 'There is an original and sterling merit in his productions that gives to all his works a peculiar stamp and character of his own for strength, clearness, and facility, without any mixture of styles', wrote Burney, and his opinion is supported to-day by Constant Lambert, to whom we are

GROWTH OF CLASSICAL ORCHESTRA

indebted for the publication of these delightful works in score and parts for present-day enthusiasts.

Burney's phrase, 'without any mixture of styles', falls strangely on modern ears, for the eighteenth-century composers varied so much in their choice of titles for instrumental music that it seems difficult to pin down any definite title to many of their compositions. Handel's Organ Concerto, Op. 7, No. 2, is described in Breitkopf and Hartel's catalogue as an overture, and Boyce's eight symphonies are closely related in form to the Italian overtures of the time. Styles were beginning to merge into classical sonata form, but they had not yet lost touch with their origins in church, chamber, or the opera house, nor had the distinction between French and Italian styles of overture ceased to exist.

The *Ouverture à la manière française* was of some importance in the development of early orchestral music. It consisted of a slow introduction, followed by a quick movement in a fugal style, which formed the main part of the overture; a final section — slow but not so slow as the introduction — completed the work. This type of overture was used well into the eighteenth century; Handel's overture to *The Messiah* is an example that springs readily to mind. Contrary to this was the Italian overture, or *Sinfonia avanti l'opera*, the three sections of which were respectively quick, slow, and quick. Preference was for a good four-square opening, and a swinging triple-time ending, with an expressive melody in between. The phrase *Sinfonia avanti l'opera*, however, need not be confined to a description of this form — it described the instrumental introduction to an opera or other work, even when this happened to be of the French type. From this a loose use of the word 'symphony' became general, to describe any independent instrumental movement, whether at the beginning or in the course of a work employing voices in other sections, or a purely instrumental composition written for concert performance.

When continuity of mood had to be maintained in instrumental music, the device employed was that of the *ritornello*,

THE ENGLISH MANNER

which, as the name implies, consisted of turning round a little melodic fragment in a variety of ways, maintaining the mood by repetitive devices. From this simple source of instrumental continuity came the technique of thematic development that made possible the structure of sonata form, and with it the classical symphony and concerto, but these terms could not be used with the same definite meanings in the pre-classical period — a symphony in the early eighteenth century might mean any instrumental composition that could not easily be described as an ayre or by the name of a dance-form.

The difference between two 'symphonists' like Boyce and J. C. Bach is that J. C. Bach shows a more advanced feeling for thematic development than Boyce. Bach makes more use of *ritornello* technique in an endeavour to create instrumental movements that are longer and freer than the simple binary and ternary forms of the dance-tunes used by Boyce. Boyce takes care to see that his dances and fugues are good of their type and arranged in such a way as to form a compound artistic scheme in which the beauties of the individual movements are enhanced by their position in the symphony. 'They have', says Constant Lambert, 'a vigour and a charm that are rarely found together', and although the late eighteenth-century taste for massive Handelian orgies in the Commemoration Festivals, continued through the nineteenth century, has had the effect of retarding the renown that is Boyce's due, those days have passed, and we are free to hear his symphonies once again in a community that will accept music as an entertainment without necessarily demanding that it should have an aura of moral uplift.

Of the eight symphonies of Boyce, No. 1 in B flat, No. 2 in A, and No. 6 in F, are scored for strings and oboes only; the oboes, as was the common usage of the time, playing at the same pitch as the violins. There is no soaring of the first violins up to the fifth position, as with Handel — technically the Boyce symphonies are simple, but they are nevertheless a delightful entertainment. All the general characteristics of the instrumentation of the time, which have been dealt with in chapter two, govern Boyce's orchestration; a study of any of his symphonies in rehearsal with the school orchestra would be, in fact, the best way of introducing the general characteristics of the pre-classical symphony to students, preparatory to a study of Haydn and Mozart; their standard of technical difficulty is sufficiently low to present no trouble to an amateur orchestra capable of playing the better-known symphonies of Haydn, and the musical qualities of Boyce's work will amply reward those who are prepared to make the experiment.

The usual treatment of wood-wind and strings is to be seen. In none of the symphonies is there much independence of these instruments except in points of imitation; oboes and violins keep together, and in the seventh of Boyce's symphonies, in B flat, and the eighth in D minor, where flutes are used, they serve as alternatives to oboes, playing with the violins, and at the same pitch, but not in the same movements as the oboes. Symphony No. 3 in C, however, employs a bassoon, which doubles the basses in the first movement — a fugal one — and the alto part in the second movement, a procedure that certainly suggests that Boyce understood the possibilities of the bassoon's upper register. It is likely that the general practice of restricting the bassoon to the bass in the eighteenth century was due less to ignorance of its tenor register than to experience of its intonation. This would be bad in the upper register of their instruments. Boyce's use of the bassoon to double the alto part — for he nowhere allows it to play solo — was risky, and may be accounted for by the fact that his symphonies were written after his deafness had become almost total. We shall do well to bear in mind, however, that Spohr used this same argument with regard to Beethoven's last quartets and Spohr was wrong. Handel's opening theme of his fourth Organ Concerto of the Op. 7 set is much more original than Boyce's bassoon part in his third Symphony, and would be even more obvious if out of tune. The explanation may be that just as Beethoven wrote 'impossible' parts for contrabasses and Berlioz wrote 'impossible' parts for trombones, on the theory that 'these notes are in the instruments and the players must get them out', so Handel and Boyce wrote music well within the compass of the bassoon, believing that a time would come when the craftsmen who made the instruments and the musicians who played them would devote their skill to a solution of the composer's problem; and that is precisely what happened, though it took nearly a hundred years to do it.

In Boyce's fourth Symphony in F, strings and oboes are used with two horns in the second and third movements; here the horns have the same melodies as the violins and oboes, but are given more opportunities for maintaining them on their own than Boyce allows to wood-wind. Symphony No. 5 in D uses two trumpets, and is consequently the most vigorous of all — the trumpet's martial character asserts itself; it decides the mood of the symphony, and the unsuitability of the strings and wood-wind to fanfare themes leads to a distinctive treatment of the trumpets and a consequent independence in their parts that Boyce has not found necessary in his other symphonies. This symphony is the longest and best-developed of Boyce's eight.

It cannot be claimed that Boyce has the same dash and daring as Handel. Boyce was writing for an assured public whereas Handel had to attract an audience to his theatre by hook or by crook. The object of Handel's organ concertos, as we have seen, was to afford relief to the strain on the less serious members of his oratorio audiences. Boyce, as master of the king's band, was in the same fortunate position as Haydn at Esterhaz; he had a band of between twenty-four and twenty-six players, and an assured salary. His duty was to provide entertainment, which his symphonies did just as effectively as his masques and pastoral pieces. The idea of his entertainments being written for posterity might have been strange to him, but the collection of cathedral music was not. It was another case of the English tradition for music with a purpose being preferred to music for mere enjoyment. There was at this time a parallel feeling for a more serious view of religion, which found expression in the Methodist movement; for, although most of the people who felt a preference for secular oratorio over instrumental music would have disclaimed any sympathy for Methodism, the causes were psychologically the same.

SUBSCRIPTION CONCERTS

THE artistic dangers inherent in the necessity of satisfying the desires of patrons of music have been the complaint of all disappointed composers and not a few musical historians. It must be admitted that no system offering complete freedom of expression and freedom from economic pressure has ever been found. Perhaps it is too much to expect that the musician should enjoy protection from those practical affairs which dominate the lives of most people; nor is it on the evidence available capable of being proved that art will flourish more in a carefree mind than in a mind obliged to accept social and economic responsibilities. Certainly the social conditions of the eighteenth century had much to do with the peculiar trend of orchestral music in England, not only during that century, but in later years.

The educational system carried out in the eighteenth century by our public schools and universities segregated the nobility, clergy, and squirearchy from the rest of English society but taught them precious little. It was completed, however, by a continental tour, which brought this small section of the community in touch with European affairs, both political and cultural. The music of the Italian and German masters was known to this class of Englishmen. When they settled back in their homes, therefore, they were prepared to encourage continental musicians to come to England or send copies of their compositions to be played here. Moreover, they knew the names of the composers most talked about on the Continent at that time, and these, and only these, they were prepared to encourage. When arranging their semi-private subscription concerts they asked for symphonies by the most modern composers like Haydn and Stamitz, but they were nevertheless sure of what they would hear. The experimental risks were taken by the patrons of such composers on the Continent. Subscription concerts therefore were cosmopolitan in artistic but insular in social outlook. The peculiar function of the foreign musician in this social scheme may perhaps be explained by this double outlook. Certainly the demand for subscription concerts increased as the century advanced:

> After the death of Festing, [writes Burney] the subscription concert at Hickford's room declined, and another was established by Mrs. Cornelys in Soho Square; where the best performers and the best company were assembled, till Bach and Abel, combining interests, opened a subscription about 1763, for a weekly concert; and as their own compositions were new and excellent, and the best performers of all kinds which our Capital can supply enlisted under their banner, this concert was better patronised and longer supported than perhaps any one had been in this country; having continued for full twenty years with uninterrupted prosperity. The same concert now subsists in a still more flourishing way than ever, under the name of The Professional Concert, with the advantage of a greater variety of composition than during the regency of Bach and Abel, to whose sole productions the whole performance of each winter was chiefly confined ... Here Cramer, Crosdil, Cervetto, and other eminent professors established their reputation, and by every new performance, mounted still higher in the favour of the public.

Had J. C. Bach kept to his orchestra the world would have been better off for his efforts, but he seems to have used the subscription concert as a sure means of obtaining a living, reserving his best energies for the opera house. It is another example of the attraction of Italian opera for the composer — one of the most unfortunate crazes of that time — for among other things J. C. Bach was a pioneer of pianoforte music, and the world is the poorer for the fact that he did not hesitate to confine his efforts at first to concertos that ladies could 'execute with little trouble'. It is in such music as this, and Handel's harpsichord music, that the justification lies for Hawkins' observation on composers who

SUBSCRIPTION CONCERTS

'like most of that profession who are to live by the favour of the public, have two styles of composition, one for their own private delight, the other for the gratification of the many'.

In the case of J. C. Bach, however, it must be remembered that he was one of London's foremost teachers, and there must necessarily be a difference between compositions written for elementary tuition and those written for concert performance by professionals. Burney can throw light on the problem; he says:

> There is no instrument so favourable to such frothy and unmeaning music as the harpsichord. Arpeggios, which lie under the fingers, and running up and down the scales of easy keys with velocity, are not difficult, on an instrument of which neither the tone nor tuning depends on the player ... And Mr. Babel, by avoiding its chief difficulties of full harmony, and dissimilar motion of the parts, at once gratified idleness and vanity ... At length, on the arrival of the late Mr. Bach, and construction of pianofortes in this country, the performers on keyboard instruments were obliged wholly to change their ground; and instead of surprising by the *seeming* labour and dexterity of execution, had the real and more useful difficulties of taste, expression, and light and shade, to encounter.

J. C. Bach's first sets of harpsichord concertos, published as Op. 1 and Op. 7, have all the features Burney so rightly condemns, but the fourth concerto of the third set, published as Op. 13, interested Haydn so much that he arranged it for pianoforte solo. In comparison with Bach's concertos Bach's symphonies, written for his subscription concerts, are much better, and it is well known that both Haydn and Mozart — especially the latter — learned from them. They have a melodic charm that needs no further recommendation. And so, although they were so much concerned with their own times and so little with posterity, J. C. Bach and his partner Abel have their places in the general history of music, and some attention must be given to their work.

GROWTH OF CLASSICAL ORCHESTRA

Karl Friedrich Abel arrived in London in 1759 and secured an appointment as chamber musician to Queen Charlotte with a salary of £200 a year. He had been born in 1725 and learned his trade as a musician under J. S. Bach at the Thomasschule in Leipzig. Before coming to London he had also spent ten years in the court band at Dresden. Abel had therefore known John Christian Bach as a boy, and the circumstances that led them to London within three years of each other can be traced to the same root in the political and economic unrest of their times. Had Frederick the Great not been involved in the Seven Years War in 1756 the petty princes of central Europe might not have been forced so soon into financial difficulties, and would have retained longer their musical establishments. J. C. Bach was with his brother Carl Philip Emanuel at the court of Frederick until the outbreak of the Seven Years War, but left for Italy in that year. At Potsdam he had had opportunities to study all kinds of music, including his brother's symphonic style, and there he became acquainted with the new pianofortes, but Italian opera was his choice. He spent six years in Italy, and it was as composer of Italian opera at the King's Theatre that Bach came to London in 1762. He sought out his father's pupil, Abel, and they lodged together. (Abel had left Dresden when the economic disturbances of the Seven Years War began to grow acute.) Soon Bach and Abel were jointly engaged in professional enterprises. In 1764 they gave a concert at Vauxhall Gardens, in 1765 they jointly conducted Mrs. Cornely's subscription concerts, and in the same year they combined with Gallini, a Swiss dancing master who enjoyed Royal favour, and had lately taken over the management of the King's Theatre, to erect a concert hall called the Hanover Square Rooms. This building remained a centre for fashionable balls and music until well into the nineteenth century. Originally Gallini had a half share and Bach and Abel each a quarter share, but Gallini bought out his two partners the year after the rooms were built. Here Bach and Abel held their subscription concerts, here the Professional Concert met, and here Salomon

SUBSCRIPTION CONCERTS

introduced Haydn to the London public in 1791. The Hanover Square Rooms, in fact, saw every important development in orchestral music in London during the precarious period of that art's establishment.

Abel's playing is said to have been facile. He played many instruments besides the bass viol, and, like Handel, was always interested in new ones. If Bach was Abel's inferior as a performer, however, he was superior as a composer. Abel's symphonies are solid substantial works, with an understanding of C. P. E. Bach's technique, but lacking his imagination. J. C. Bach's symphonies, on the other hand, are his own; he has the German liking for thematic development (as far as it went with the younger Bachs), and a sunny smile in his melody that may perhaps suggest the reason why he leaned so much towards the Italians in his youth. Some of Bach's symphonies, concertos, and overtures are available in modern editions, that of Adam Carse being admirable. His orchestra usually calls for no more than a pair each of oboes and horns in addition to the usual strings and continuo, but Bach was by no means limited to these slight forces throughout his London career; two Grand Overtures, one for a single and the other for a double orchestra, make use of a double set of oboes, flutes, clarinets, and horns in addition to strings. They are overtures in the Italian style, beginning and ending quickly, but with a contrasting slow middle section.

Among those works made available for modern players by Adam Carse, J. C. Bach's *Symphony in B flat* should be compared with C. F. Abel's *Symphony in E flat*. Both are well within the technical limitations of a good school orchestra, and indeed there is no reason why amateur orchestras in some of our smaller towns, where large-scale orchestral concerts are impracticable owing to the lack of a suitable hall, should not obtain excellent renderings of some of these pre-classical symphonies. J. C. Bach's *Overture in B flat*, also available in the edition by Adam Carse, could be similarly utilized.

It must be emphasized, however, that, unlike some of his

GROWTH OF CLASSICAL ORCHESTRA

harpsichord and pianoforte concertos, J. C. Bach's overtures and symphonies were not composed as teaching material: they were intended as concert works. Within their conventional three-movement form he was able to adapt himself to the various needs of his audiences. The overtures are slighter works than the symphonies, and the symphonies are less imposing than the more heavily-scored works that he called Grand Overtures, Sinfonia Concertante, or Concert ou Symphonie. These latter were probably influenced by the size of the new Hanover Square Rooms, for the large orchestra they demand would have been out of place in the more intimate but restricted space of Carlisle House or Almack's Room. A further indication of the social background of the symphonies is the occasional use of a minuet for the finale. Boyce's use of a Gavotte to conclude his *Symphony in D minor* is another example of the same tendency. The time had not yet come, however, for a dance movement to establish itself as the third movement of a four-fold symphonic scheme such as we are now accustomed to hear, nor has such a movement even now become part of the solo concerto convention.

Although Bach and Abel's subscription concerts were popular they were not the most important socially in the London of George III. 1776 saw the inauguration of the Concert of Antient Music with a most imposing list of directors. The title was generally shortened in speech to the Ancient Concerts, and they used to meet first in the Tottenham Street Rooms, which later became the Queen's Theatre in Tottenham Court Road. Their first orchestra consisted of forty-three players, made up of 16 violins, 5 violas, 4 'cellos, 4 oboes, 4 bassoons, 2 double basses, 2 trumpets, 4 horns, 1 trombone and 1 drum. The presence of the trombone suggests that permission had been obtained to employ one of the players from the king's band, and indeed this is likely to have happened, for in later years the king came to be something more than a nominal patron; in 1785 George III became a regular attender at these meetings, accompanied by Queen Charlotte and those of the princesses

SUBSCRIPTION CONCERTS

old enough to appear in public; the king is reputed to have written out the programmes himself, and the organization was in consequence sometimes called the King's Concert. The musical policy of the Concert of Antient Music was conservative; it was their rule to play only works with an established reputation, and all compositions dating from less than twenty years back were barred from their platform. This policy continued throughout the society's seventy-two years of existence, and is of some importance in the study of orchestral policy in England. The Ancient Concerts did good work in keeping before the public good music that might have been neglected had the modernist had things all his own way — they included madrigals in their programmes at times when these compositions were regarded by many musicians as obsolete, and they gave something more than lip service to the works of Henry Purcell. Bach and Abel were foreigners interested in performances of their own works and those of their continental friends who likewise composed in the new symphonic style — it was no part of their business to propagate English music of types believed to be declining in favour — there was, then, a place for a conservative concert-giving society in London, and musical life would have been poorer in the capital had the Concert of Antient Music not existed.

But since the twenty years rule made it impossible for the modern composer to approach the London public through the Ancient Concerts, had Bach, Abel, and the Professional Concert not existed the new symphony and its orchestra would have had little place in London life, for the orchestral style of the Ancient Concerts was that of Handel. Many London people, as we have seen, were interested in the latest developments in orchestral music abroad, and many players from famous continental orchestras had come to London. Wilhelm Cramer, already mentioned as the leader of the orchestra at the Handel Commemoration of 1784, had had the benefit of tuition from the elder Stamitz of Mannheim (being a son of one of Stamitz's violinists), and was admitted to the Mannheim orchestra at the age of sixteen. This orchestra is well known as that in which Mozart first heard clarinets, and Stamitz was one of the first to open up the possibility of a wider range of dynamics by the exploitation of crescendo and diminuendo effects. Loud passages alternating with soft passages had been generally sufficient before that time. The symphonies of the two Stamitzes — father and son — were heard in London and published in orchestral parts. Wilhelm Cramer came to London in 1772, became head of the king's band, leader of the orchestra at the Pantheon opera house and of the Ancient Concerts, he frequently appeared at Bach and Abel's concerts, and J. C. Bach thought so well of Cramer's master Stamitz, that, on his only visit to the Continent after taking up his residence in London, he called on Stamitz at Mannheim. History has not been kind to Stamitz's compositions, but his London contemporaries took an interest in his pioneer work.

Haydn's name appears for the first time in England on Bach and Abel's programmes. It is not therefore surprising that by 1783, when Bach was dead and Abel had temporarily left England, the public missed this contact with modern developments in orchestral music. The section of musical society most affected was professional — especially those foreign instrumentalists who had made London their home. So the Professional Concert came into being. This organization existed to supply the demand for new compositions for the concert-hall ruled out of the programmes of the Concert of Antient Music. The new organization had not anything like the same social prestige as the Ancient Concerts, however, which was an amateur-controlled body disdaining any suggestion that it was interested in making money out of music.

The Professional Concert had nothing to fear, however, from the Ancient Concerts; there was room for both societies because they existed to supply different standards of taste. Opposition came from within. Among the German musicians who had lately been attracted to London was Johann Peter Salomon, who came to London in 1781, appearing as leader and soloist

SUBSCRIPTION CONCERTS

in that year at Covent Garden Theatre. His story was similar to that of many other foreign musicians who came to England in the eighteenth century. Salomon was thrown out of work when Prince Henry of Prussia decided to dismiss his orchestra at Rheinsberg, where Salomon was konzertmeister. Salomon travelled first to Paris, where his playing was much appreciated, but he nevertheless decided not to settle there but to move on to London. Here his playing was found to be superior to Wilhelm Cramer's, and his position seemed secure; he joined the Professional Concert on its formation, but quarrelled with them in the first year and was in consequence debarred from their platform. His reaction was the same as that of John Banister, Handel, and many other musicians of enterprise — he started a business of his own. Salomon's concerts aimed to do the same work as the Professional Concert but to do it better. Within a short time the Professional Concert began to lose money.

This was due not only to Salomon's violin playing, fine as it was, but to his personality and artistic policy. His manners made a good impression from the first, and in later years his unimpeachable honesty increased his prestige. He had good judgment, too, in the trend of orchestral music; he saw that Haydn, in the seclusion of the Esterhazy country seat, had at last gained a mastery of that new symphonic form with which so many of his contemporaries had been struggling. Salomon had lost no opportunity of performing Haydn's works wherever he had been. In London he found Haydn in demand, and, being less concerned with the performance of his own compositions than Bach and Abel had been, was moved to deal more wholeheartedly with the demand for Haydn than they.

His business rivals soon tried to outwit Salomon by endeavouring to fetch Haydn to London. Wilhelm Cramer tried in 1787 to induce Haydn to appear at the Professional Concert, offering to pay Haydn his own terms. Salomon at once sent Bland, a London music publisher, to Haydn to try and secure his services for the Salomon concerts, and an amusing story has been told of their meeting. Haydn was attempting to shave when Bland was shown into the room, and complaining bitterly of a blunt razor. 'I would give my best quartet', he snapped, 'for a good razor.' This was no time for an attempt to talk business with Haydn, so Bland staked everything on a stroke of humour. He returned to his lodgings and fetched a razor. This he presented to Haydn with a reminder of his rash offer. He returned to London with the quartet—called the *Raziermesser* or Razor Quartet — but without Haydn. Salomon's personal appeal through Bland had had no more effect than Cramer's offer of all the money Haydn chose to name. Haydn had lived a quiet, frugal life; he had not been able to save enough to provide for the time when old age should come upon him; notwithstanding this, he would not leave his master.

So the situation remained for another three years. Then Prince Nicolaus Esterhazy — 'Nicolaus the Magnificent' — died. He was succeeded by Prince Anton who saw, as so many other continental princes had been forced to see, that the *ancien régime* was heading for disaster — France, indeed, had been found to be bankrupt as early as 1787, and the revolution of 1789 sounded a warning even as far as Esterhaz — so Anton decided on a policy of economy. He dismissed nearly all his musical staff, keeping only a small number for the church services, and Haydn found himself still nominally in employment but with little to do. William Forster, the London violin maker and publisher, had issued no less than eighty-two of his symphonies, besides twenty-four quartets, twenty-four solos, duets and trios, and the church music for the service of *The Seven Last Words of Our Saviour on the Cross*. If Haydn wanted money, here was a wealthy city, London, where his works were in demand. Salomon happened to be in Cologne, returning from an expedition to Italy in search of opera singers, when he heard of the death of Prince Nicolaus Esterhazy. He heard also that the Earl of Abingdon was trying to engage Haydn to appear in London at the Professional Concert. Salomon

SUBSCRIPTION CONCERTS

immediately set out for Vienna to try his own powers of persuasion with Haydn. At any cost he would bring this prize to London.

In due course he came into Haydn's presence. 'My name is Salomon,' he announced bluntly, 'I have come from London to fetch you. We will settle terms to-morrow.'

Haydn capitulated.

HAYDN IN LONDON

When Haydn arrived in London in 1791, he stepped out of an environment where he had been a superior kind of domestic servant into one where he was a 'good commercial risk'. All the familiar forces of competitive business were brought to bear by his employer Salomon on his potential value as a popular composer of the best type. He was advertised in the newspapers, overwhelmed with social introductions, and accepted into learned associations with honour. His personal reactions to this strange life have been related by numerous biographers; they show him to be a man of simple tastes and simple honesty, seeking to escape from the noise of London streets and the distractions of innumerable social functions to the seclusion necessary for his work of composition, but drawn back again constantly by his associates in order to satisfy the public demands for his appearance.

It was not all unbiased, this honour paid to Haydn; Salomon had agreed to pay him £50 for each of twenty performances, and had to make a profit for himself after defraying all other expenses. In addition, Haydn was to have the proceeds of two benefit concerts at each of which £200 was guaranteed to him. It was not to be expected that business rivals would make Salomon's task an easy one, yet the course of events shows that the fight was decided by a conflict of artistic and social forces rather than by purely financial interests.

London's musical supporters were divided into two groups — the conservative and the progressive. The former centred round the Concert of Antient Music and the Italian opera, which had now been transferred to the Pantheon, after the destruction by fire of the King's Theatre in 1789; the progressive faction centred round the Professional Concert, and Salomon. Gallini, who had tried to persuade Haydn to write an opera for a new opera house he was to open in the Haymarket, came into the fight as a business competitor of Salomon, involved willy-nilly in the social and artistic complications of the affair, but having to make the best bargain he could in his own financial interests.

King George III was a staunch supporter of the Concert of Antient Music; the king, too, held the view that a second opera house was unnecessary, so the Lord Chamberlain refused Gallini a licence. This in turn frustrated Salomon's plans, for he had engaged two of Gallini's vocalists, Cappelletti and David, for his first Haydn concert. Cappelletti and David were under contract to Gallini not to sing in public before the opening of the new opera house, and Gallini held them at first to this contract. Salomon had therefore to postpone Haydn's first symphony concert until these singers should be available. Meanwhile Salomon's opponents made the most of the delay. The newspapers jibed at German musicians who came to this country with a great flourish of trumpets to 'charm the money out of the pockets of John Bull'. They did not hesitate to suggest that Haydn had met with little recognition in his own country, and would probably prove inferior to such players as Cramer and Clementi. Gallini, finding himself opposed by Salomon's enemies, made common cause with him; he applied for a licence for 'entertainments of music and dancing' instead of opera, released David from his contract so that he could appear on March 11th 'whether the Opera House was open or not', and engaged Haydn, Salomon, and his orchestra to appear at concerts in his new premises. So, after much delay, Haydn was allowed to prove his worth to the public.

Salomon's orchestra for the Haydn concerts was of good strength, varying in size from thirty-five to forty players, led by Salomon himself, with Haydn presiding at the keyboard. This orchestra, playing in the Hanover Square Rooms, which measured ninety-five feet by thirty-five feet, was the largest Haydn had ever had at his disposal.[1] The opening concert used an

---

[1] Larger orchestras had played Haydn's symphonies, e.g. the 'Oxford Symphony was written for the Concert Spirituel (60 players) but Haydn did n conduct it in Paris.

orchestra of 16 violins, 4 violas, 3 'cellos, 4 basses, flutes, oboes, bassoons, trumpets, and drums, for the Symphony in D, No. 93, which was enthusiastically received, and the slow movement encored, greatly to Haydn's satisfaction, for such an honour was rarely given to an instrumental movement.

There was good reason for the honour. Apart from the merit of the symphony, there was the quality of its performance, which Haydn had striven to bring up to the standard of his own orchestra at Esterhaz. Whether he did this or not will never be known, but Dies records in his *Biographische Nachrichten von Joseph Haydn* how the composer behaved at his first rehearsal with the Salomon orchestra. The first three notes were played much too loudly for Haydn, who promptly stopped the orchestra and called for less tone. Three times he did this without getting a satisfactory result. Then Haydn heard a German player whisper in his own language to his neighbour: 'If the first three notes don't please him, how shall we get through all the rest?' Haydn gave up trying to explain in speech, borrowed a violin, and demonstrated the tone he wanted to be produced. After that he had no more trouble with the passage. This was not the only occasion on which Haydn had to demonstrate to an English orchestra how a passage should be played; possibly he had to do so often, for a second authentic record has been related by Sir George Smart, who was engaged as a violinist at one of Salomon's later concerts. Young Smart[1] was then at the beginning of his instrumental career, and anxious to please Haydn; so much so that he volunteered to play the drums in the absence of the player engaged. Since Smart had never previously attempted to play a drum it follows that Haydn had some agonizing moments when this new broom got into action. At last, taking the drumsticks out of Smart's hands, he played the passage for him, saying afterwards:

'That is how we use the drumsticks in Germany.'

'Oh, very well', replied the cocksure fiddler, 'if you like it better that way we can also do it so in London.'

[1] Smart was knighted in 1811.

HAYDN IN LONDON

Such experiences must have puzzled Haydn at first, but throughout he showed himself not only a capable, but a tactful conductor. These men, he knew, lacked discipline, for they came together in all sorts of situations under different leaders and had to adapt themselves quickly to unusual demands. They had sharper wits than his permanent continental orchestra, but less opportunity for continuous rehearsal and for developing a team spirit. Haydn's styles in composition and performance were fundamentally opposed to many of the performances in which these players were required in London to take part, for Haydn had learned the method of placing his instruments in a *tutti* chord so that the full effect of the overtones could be heard: Salomon's forty players under the direction of Haydn playing one of that master's compositions were more effective than a hundred would be playing music in the Handelian style — and Handelian style was still very much to the fore in London musical circles, for the King's Concerts (as the meetings of the Concert of Antient Music were called) relied more on Handel's music than on any other.

Salomon was, as we have seen, opposed both artistically and financially to this influential Society; how great a difference lay between their ideals can be seen by a comparison of Haydn's Salomon symphonies with the most fashionable musical festivals of the time — the Handel Commemoration performances in Westminster Abbey and the Pantheon, the first of which took place in 1784 and the last of which in the eighteenth century Haydn himself attended in 1791, occupying a seat of honour near the royal box. There were 885 people engaged in the performance, arranged on specially-built tiers reaching right up to the roof of the Abbey. The weight of tone was so impressive that during the singing of the Hallelujah Chorus Haydn wept like a child: 'He is the master of us all', he said.

But the Handel Commemoration performances, nevertheless, left much to be desired. The principles of orchestration that Haydn had spent his life helping to evolve were not in evidence; here the prevailing idea was to employ as many instruments

GROWTH OF CLASSICAL ORCHESTRA

and voices as possible in the belief that these, and not the scoring, were the deciding factors in musical effect. This is clear enough from Burney's *Account of the Musical Performances in Westminster Abbey and the Pantheon in Commemoration of Handel,* descriptive of the 1784 festival, where he says:

> In order to render the band as powerful and complete as possible, it was determined to employ every species of instrument that was capable of producing grand effects in a great orchestra[1] and spacious building.

With this object in view the Concert of Antient Music (which society had been entrusted with the arrangement of the performances) gathered together a total of 525 musicians — singers and players — comprising 59 sopranos, 48 altos, 83 tenors, and 84 basses; 48 first violins, 47 second violins, 26 violas, 21 'cellos, 15 double basses, 6 flutes, 26 oboes, 26 bassoons, 1 double bassoon, 12 trumpets, 12 horns, 6 trombones, kettle drums, tower drums, and organ. It will be seen from the proportions of this band, that the directors were still acting on the conception of orchestral balance common in the first half of the eighteenth century. There are roughly two oboes to seven violins, which is somewhat less than the earlier proportion of oboes to violins, but there are only six flutes to twenty-six oboes, which seems to indicate that the flute still occupied an inferior position in London concert life; the bassoons outnumber the 'cellos, and even with the string basses and the organ, the preponderance of their tone must have been obvious, especially as the double bassoon was placed in a proud position in front of the conductor. (It is printed in big letters on Burney's plan of the disposition of the orchestra.) Burney was particularly proud of the pains that had been taken to collect such new instruments, and whenever possible gives Handel credit for having first introduced them, as also of the method of playing the organ by remote control from the harpsichord keyboard. Handel,

[1] Burney is here using the word 'orchestra' in the Greek sense, to mean the place occupied by the performers: in this case the whole of the West end of the Abbey nave.

as we know, preferred to play his *continuo* on an organ, but a harpsichord had the advantage for a conductor that his hands could more easily be seen by the players grouped round the instrument. According to Burney, Handel had the keyboard of a harpsichord connected by wires to the keyboard of an organ at some of his oratorio performances, thereby keeping better control of his forces while having at the same time the sustained effect of organ tone. This system was adopted at the Commemoration, the wires extending a distance of nineteen feet from one instrument to the other. As Burney truly observes, 'to convey them so great a distance from the instrument without rendering the touch impracticably heavy, required uncommon ingenuity and mechanical resource'.

From the harpsichord keyboard the conductor at the Handel Commemoration had immediate direction of the various leaders arranged before him, with the principal first violin in a specially prominent position. Altogether twenty-eight principals were employed: 2 for first violins, 2 for second violins, 4 for violas, 4 for 'cellos, 4 for double basses, 4 for oboes, and 4 for trumpets; a complicated arrangement in all conscience. But Burney is quite proud of it, and from it strives to prove how superior London orchestras were over those of Paris, where, both at the Concert Spirituel and the Opera it was usual to conduct with a baton. Burney says:

> Foreigners, particularly the French, must be much astonished at so numerous a band moving in such exact measure, without the assistance of a *Coryphaeus* to beat the time, either with a roll of paper, or a noisy baton, or truncheon. Rousseau says that 'the more time is beaten the less it is kept' and it is certain that when the measure is broken, the fury of the musician-general, or director, increasing with the disobedience and confusion of his troops, he becomes more violent, and his strokes and gesticulations more ridiculous, in proportion to their disorder.
> The celebrated Lulli, whose favour in France, during the last century, was equal to that of Handel in England, during

GROWTH OF CLASSICAL ORCHESTRA

the present, may be said to have beat himself to death, by intemperate passion in marking the measure to an undisciplined band: for in regulating, with a cane, the time of a *Te Deum*, which he had composed for the recovery of his royal patron, Louis XIV, from a dangerous sickness, in 1686, he wounded his foot by accidentally striking on that instead of the floor, in so violent a manner, that, from the contusion occasioned by the blow, a mortification ensued, which cost him his life, at the age of fifty-four!

And, as the power of gravity and attraction of bodies is proportioned to their mass and density, so it seems as if the magnitude of the band had commanded and impelled adhesion and obedience, beyond that of any inferior force. The pulsations in every limb, the ramifications of veins and arteries in an animal could not be more reciprocal, isochronous, and under the regulation of the heart, than the members of this body of musicians under that of the Conductor and Leader. The totality of sound seemed to proceed from one voice, and one instrument; and its powers produced, not only new and exquisite sensations in judges and lovers of the art, but were felt by those who never received pleasure from Music before.

These effects, which will be long remembered by the present public, perhaps to the disadvantage of all other choral performances, run the risk of being doubted by all but those who heard them, and the present description of being pronounced fabulous if it should survive the present generation.

Such confidence had been sufficient to inspire four repetitions of the Commemoration by the time Haydn came to hear it in 1791. The only difference between that recorded by Burney and that heard by Haydn being that the number of performers had increased from 525 to 885. No doubt Haydn heard opinions very similar to Burney's expressed, too, on the merit of the players:

Few circumstances will, perhaps, more astonish veteran musicians than to be informed that there was but *one general rehearsal* for each day's performance: an indisputable proof of the high state of cultivation to which practical music is at present arrived in this country; for, if good performers had not been found, ready made, a *dozen rehearsals* would not have been sufficient to make them so.

But Haydn had nevertheless to cope with such situations in rehearsal as are recorded by Dies and Smart. He found himself, in fact, in the midst of a somewhat backward musical environment, but respected as a composer of the highest merit. His music was admired, and if at first there had been some influential support of Salomon's business rivals, with their derogatory remarks on Haydn's motives and ability, his almost complete lack of a similar spirit in retaliation, and the electrifying effect of his performances, rendered such slanders singularly harmless.[1] Where Handel had been driven to the verge of apoplexy Haydn's sense of humour won him friends: the *'Surprise' Symphony* is a case in point. 'There all the women will scream', said Haydn, pointing to the *paukenschlag* — the explosive *fortissimo* on the drum, after the sleepy opening of the slow movement — but he knew that they would like it. That was his business — to entertain, not to reform — and in the process of entertainment a pleasant novelty is never amiss. It was all very different from the Handel Commemoration orchestration; Haydn had perfected the scheme of orchestration that was to be the glory of the classical symphony: the harpsichord or organ was no longer necessary to fill in the harmonic scheme — the Handelian method of scoring was (except among the backward) a thing of the past. As Mr. Frank Howes says:[2]

> This method of scoring became completely outmoded in the course of the quarter century between Bach and Haydn. Horns and oboes are capable of sustaining harmony between treble and bass even better than the harpsichord, and it only needed to brighten their tones with flutes and to

---
[1] On one occasion only does Haydn seem to have derided a competitor: he said that Giardini 'played like a pig'.
[2] Frank Howes, *Full Orchestra*.

GROWTH OF CLASSICAL ORCHESTRA

> mellow them with clarinets to have at the composer's disposal a chorus of wind instruments, which he could use, not in the manner of an obbligato, but all together either with or without the strings to sustain the harmony, to reinforce the tone and to exploit the newly discovered powers of tonal graduation by crescendo and diminuendos.

There was nothing in this improved method of instrumentation that need offend the ears of a conservative listener, even if J. C. Bach and Abel had not prepared the way for it. It is ludicrous to think how the Professional Concert, which had carried on the work of Bach and Abel, tried to discountenance Haydn by bringing over his pupil Pleyel as a counterblast. True, they only did this after every attempt to coax Haydn into breaking his contract with Salomon and Gallini had failed. Haydn's honesty, as well as his musical ability, endeared him to London society. His triumphs even came to be reported on the continent, for the *Journal of Luxury and Fashion*, published at Weimar in 1794, contains a London notice which reads:

> But what would you now say to his new symphonies composed expressly for these concerts, and directed by himself at the piano?[1] It is truly wonderful what sublime and august thoughts this master weaves into his works. Passages often occur which render it impossible to listen to them without becoming excited. We are altogether carried away by admiration, and forced to applaud with hand and mouth. This is especially the case with Frenchmen, of whom we have so many that all public places are filled with them. You know that they have great sensibility, and cannot restrain their transports, so that in the midst of the finest passages in soft adagios they clap their hands in loud applause and thus mar the effect. In every symphony of Haydn the adagio or andante is sure to be repeated each time, after the most vehement encores. The worthy Haydn, whose personal acquaintance I highly value, conducts

---

[1] Between Haydn's 1791 and 1794 visits the piano had supplanted the harpsichord at Salomon's concerts.

himself on these occasions in the most modest manner. He is indeed a good-hearted, candid, honest man, esteemed and beloved by all.

But although Haydn's simple honesty had the effect of disarming Salomon's business adversaries and, finally, of bringing their patrons round to his side (this was true even of George III), Haydn's reputation would not have stood so high had his music not made an irresistible appeal to those who heard it. Salomon's public had been the usual clique of seekers after fashionable entertainment, similar in every way to Bach and Abel's public; there had been no indication that this public wanted, or would tolerate, any depth of emotion in their music. The lesson of the *Surprise Symphony* is that Haydn himself found them so, for when it was performed at his second concert in London, a newspaper critic, as if to confirm Haydn's view that the London music-lovers needed a lesson, naïvely produced a pastoral programme for it. 'The *Surprise* might not be inaptly likened to the situation of a beautiful shepherdess who, lulled to slumber by the murmur of a distant waterfall, starts alarmed at the unexpected firing of a fowling-piece.' But such shallow criticism began to find itself lost in the depths of later criticism. The passage quoted from the Weimar *Journal of Luxury and Fashion* suggests that an aesthetic revolution was even in 1794 taking place, for when it says, 'It is truly wonderful what sublime and august thoughts this master weaves into his works', it proves that Haydn had succeeded in his efforts to get his slow movements taken seriously. 'Passages often occur which render it impossible to listen to them without becoming excited.' This in the age of reason, the Augustan age, the age when formal beauty was said to be the rule, and emotionalism frowned upon by everyone who had any claims to breeding and education. Possibly some lingering shame of their show of enthusiasm still remained, so the Frenchmen (who would no doubt be refugees from the 1792 revolution) were blamed, but we have already seen how Handel's oratorios came to be preferred to his more formal works some forty years before Haydn came to London.

GROWTH OF CLASSICAL ORCHESTRA

It was a sign of the emergence of something essential to the English character; something which had led Anthony Wood's friends to prefer viols and fantasias to violins and country dances; something that had led Samuel Pepys to disapprove of Charles II's light and airy taste in church music; something that had led Boyce to set more store by his collection of cathedral music than by his own pastorals and symphonies. But it is significant that all the works of emotional depth had been written in the old contrapuntal manner; the new homophonic manner, with its experiments in form and colouring, had a long way to go before it became capable of anything deeper than a novel evening's entertainment, even if anything more had been demanded of it by the private patrons whose support was essential to its development on the Continent.

When in the late eighties of the eighteenth century symphonic form became, in the hands of Haydn and Mozart, capable of aesthetical expressions of a deeper and more varied range, there arose the problem of finding appreciative patrons to accept and further the establishment of these greater compositions. Mozart found real difficulty here. He had broken off his relations with his patron ostensibly because the latter treated him with contempt, but really because the Archbishop of Salzburg's contempt found expression in frustrating Mozart's artistic self-expression. Mozart sought the wider public of the principal cities of east central Europe, and even approached a new source of patronage through the companionship of Freemasonry, only to be overwhelmed by the magnitude of the task. It was beyond the means of any single composer to reform the social scene and fit it for worthier music; the old system of private patronage had served its purpose, the demand was now for the expansion of orchestral resources and the expansion of ideas. Haydn and Mozart found some support for the profounder work which began to issue from their minds round about 1787, but not enough. Mozart fretted under the strain; Haydn was more submissive to his customers' requirements but needed a new shop-window. Salomon provided it.

HAYDN IN LONDON

The London orchestra of Salomon provided Haydn with the means to express adequately the grander orchestral effects he and Mozart were capable at that time of realizing. Salomon's business enterprise made the performance of Haydn's twelve greatest symphonies possible in London, and the public's enthusiastic reception of these symphonies rewarded Salomon's enterprise, endowed Haydn with a competence for life, and established in this country a standard of symphonic musical expression that eclipsed all previous essays in orchestral music. This was the true starting-point of the history of the modern orchestra — the most noble and most fluent musical instrument that human ingenuity has yet evolved.

We have seen that the London public was conservative, but well informed of the new music being produced in such abundance in the courts of continental princelings. London chose the best of this music for its own concert-rooms, and London publishers were not backward in issuing vast quantities of it in print, but only that which was most talked of abroad, and well reputed. Stamitz and Haydn were in demand, but J. S. Bach was not; when Burney regretted that J. S. Bach did not 'extend his fame by simplifying his style more to the level of his judges', he spoke with the voice of his age; an age that respected Latin conciseness and disdained mediaeval complexity. Bach still leaned to the old polyphonic style that the new cult of classicism tended to reject; his hope of favour in London might have lain with the ultra-conservative Concert of Antient Music, but that society was too concerned with the cult of Handel to seek out the works of his great contemporary. On the other side was the London modernist, seeking to be *au fait* with the compositions of the most talked-of concert-masters abroad. He was prepared to spend money on keeping up to date with modern music, but he was not prepared to spend money on experimental works in the new idioms of the symphony and the concerto. This he left to continental patrons.

The cost of maintaining a private orchestra and a composer able to produce up-to-date music on request was considerable.

GROWTH OF CLASSICAL ORCHESTRA

So long as Haydn and Mozart were experimenting with strings, harpsichord, two oboes and two horns, the resources at their disposal were ample, and Haydn was exceptionally fortunate under Prince Nicolaus Esterhazy, for he was able to add flutes, trumpets, and drums, bringing his orchestra up to a total of twenty-six players. The time came, however, when the technique of orchestration reached maturity under these masters; the harpsichord was no longer necessary to hold together the harmonic structure of an orchestral composition, for the full choir of strings was balanced by a full choir of wood-wind and brass.

It was necessary for those responsible for the provision of the orchestra to acquiesce in this new point of view. It meant that the composer and not the concert promoter was deciding on the amount to be spent on orchestral personnel. To the autonomous aristocratic patron of the *ancien régime* this would appear to be equivalent to the piper calling the tune, but to Salomon it was not so. His policy was to capitalize the genius of Haydn, and in furtherance of this policy he was wise enough to give Haydn his head. Haydn in London had the orchestra he desired, which varied slightly for different symphonies; only in four of the Salomon symphonies did Haydn require clarinets — in No. 101 in D minor, No. 103 in E flat, No. 104 in D minor. In No. 100, the *Military Symphony*, he required the unusual additions to the orchestra of a bass drum, triangle, and cymbals, and used clarinets in one movement only. Here we see the full extent of Salomon's acquiescence, but the principle was accepted — the composer was free to call for the particular orchestra his musical thoughts required, and from that time onwards this principle was in evidence in all progressive compositions for the orchestra.

Haydn's contribution to symphonic progress lay in his flexibility of expression. The use of wood-wind instruments was at last freed from the conventional splitting up of forces into *concertino* and *ripieno*, as they had been in the *concerti grossi*. Now the instruments intermingled in ever-varying proportions, act-

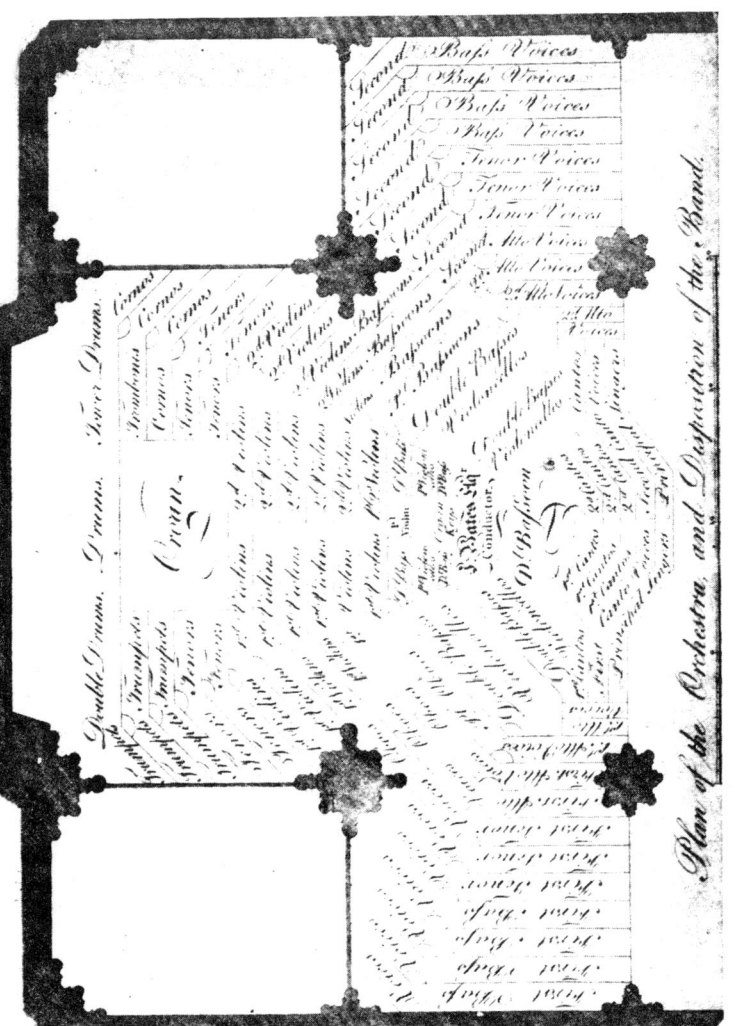

From Burney's Account of the Musical Performances ... in 1784 in Commemoration of Handel

THE B.B.C. SYMPHONY ORCHESTRA, 1942

ing sometimes as soloists and the next moment blending with the others in the instrumental choir. The long singing style of Haydn's slow movements, ornamented in a style that relied on variation of solo tone-colours far more than on the flexibility of the players' digital technique, was the feature that attracted most the attention of the Londoners, but later admirers have thought more of Haydn's spirited rustic finales, his harmonic surprises and his transformation of the stately minuet into the jocular scherzo. The twelve Salomon symphonies are the foundation of the popular modern conception of a Haydn symphony: they, almost alone of his symphonies, are remembered.[1] Yet their superiority over his earlier works in this form is so marked that the decline of his apprentice and journeyman efforts before the splendour of his master works is no cause for surprise. In them and the last symphonies of Mozart the glory of the eighteenth century shone at its brightest. The urge for formal perfection had been satisfied, but in the moment of this satisfaction a new need had become evident. It had been there all the time, but the intellectual fashions of an 'age of reason' had obscured the end to which their search for formal perfection was aimed. Only in his efforts at setting Scottish songs did Haydn really come face to face with the dawning spirit of romanticism, but unconsciously the spirit of the new age was present in Haydn's orchestral music as well as Burns' poetry. It comes out in his rusticity, the feature which posterity has recognized as being most characteristic of the true Haydn, but which London society does not seem immediately to have recognized.

The reason is evident enough. London society had almost lost sympathy with rural life. There was a spirit of selfishness abroad to which the cult of charity was a kind of moral narcotic. Rural workers were being driven more and more to pauperism while their landlords grew rich. Great estates spread over the counties and great town residences were being built in London, and it was upon the support of their owners

[1] And, of course, the 'Oxford' Symphony.

GROWTH OF CLASSICAL ORCHESTRA

that subscription concerts had to rely. Everybody of assured position in society was endeavouring to obtain or increase his revenue from the state coffers, but nobody was omnipotent. Society, even in its most corrupt period in the eighteenth century in England, had to combine into fairly large groups for musical entertainment, and these groups were the markets to which professional musicians brought their wares. Nothing was further from any of their minds than that society would soon change its face under the influence of the new controllers of industry. When Haydn came to London the influence of this new manufacturing class had not yet begun noticeably to appear there; to Tories and Whigs alike the social system rested on a custom and tradition of greatness of which the landowners were the necessary guardians. 'The age of reason' was a pleasant and somewhat flattering idea, but in England states and societies were believed not to be made but to grow, and nobody had any right to lay impious hands on them and seek to reform them in the light of reason. That, said the English, had been done in France, and look at the result. It was the sacred duty of each generation to hand on to the next the precious social heritage which had come down to it; custom and tradition would secure that this heritage should be kept intact.

This point of view was more than a political creed; it was inherent in the English way of life; the political system that could swing to a change of monarch without throwing out of gear the parliamentary machine came as a long result of this English way of thinking. In the eighteenth century the machine was being driven very cautiously because it had had a bad breakdown in the seventeenth century. In music, however, there was nothing to fear, and the English tradition, at once conservative and curious, could be allowed to operate without conscious resort to any ideology. By its curiosity we were led to encourage foreign musicians to visit our land; by its conservatism we were inclined to accept only those who were regarded most highly on the continent. In the course of encouraging these worthy foreigners, however, an English tradition of

HAYDN IN LONDON

which we were not consciously aware had its influence on our musical taste; this was the spirit that had animated our greatest men of art — Chaucer, Shakespeare, and Milton — it was in its essence romantic. 'The age of reason' as applied to eighteenth century thought was a form of artificial ideology, contrary to our customs and tradition. To foreign critics the English spirit looked like hypocrisy, with statesmen talking of liberty and denying the right of representation to those who lived in the American colonies, and emotional theologians tramping the country preaching the necessity of method in our lives. The truth is that humanity would keep creeping into the Englishman's logic like cheerfulness into Dr. Johnson's friend's philosophy. In music we applauded formal beauty but took to our hearts the sentimental ballad and the secular oratorio. Handel we loved for his serious emotional works, and Haydn for the tender feeling of his symphonic slow movements. Soon the romantic movement was about to spread over all Europe, and England would encourage its development, but not (strange as it may seem) as something inherent in her own traditions, but as an exotic foreign novelty, received by right of purchase, and copied conscientiously by a host of adoring aspirants to fame.

BOOK TWO

THE PHILHARMONIC PERIOD

THE FOUNDING OF THE PHILHARMONIC SOCIETY OF LONDON

The beginning of the nineteenth century saw orchestral music in England declining. It was a transitional period, with a large section of the population drifting from the country to the towns in response to the changes wrought by the Industrial Revolution, and the landed aristocracy no longer completely dominating the political arena as they had done in the past, but compelled to recognize the power of a new moneyed class grown influential by reason of its control of production in industry and the commercial exploitation of manufactured goods. There was a desire among these new industrialists to enjoy a reputation for charity, and, in much the same way, a reputation as patrons of music. The nineteenth century saw the upper-class resistance to these wealthy middle-class ambitions steadily being overcome until at last the position was changed, and music found its most effective supporters in the middle classes. The aristocracy had become mere figureheads in the world of music by 1880, by which date the local magnates of all our large industrial centres were actively supporting musical festivals in their own districts. It was a century of decentralization, of growing local pride, yet only in Manchester did orchestral music attain a standard comparable with that of the metropolis.

The year 1801 saw concert life in London outwardly the same as it had been in the eighteenth century, except that there was a dearth of exceptional talent. Haydn's *Creation*, performed in 1800 at Covent Garden, kept the lamp of his genius bright in London memory; indeed it did more, it associated the famous Austrian composer with the secular oratorio tradition established by Handel, so that this form of music gained greater popularity among the aristocracy, and the middle classes — who had had less to do with the subscription concerts of the last fifty years than the upper classes — added Haydn to their

THE PHILHARMONIC PERIOD

musical hierarchy as second in importance only to Handel.

The Concert of Antient Music kept up its activities and still remained the most exclusive musical organization in London. The rule was still observed that no composition written less than twenty years previously should be played at their concerts. The Royal Italian Opera relied on aristocratic patronage, but had not yet arrived at acceptance of German operas. The Lenten oratorios furnished a novelty during the theatrical close season, but their supposed sacred nature was declining before a growing taste for secular trivialities. The orchestral music at these events was, according to contemporary accounts, extremely casual; that of the opera house efficient only for the flimsy accompaniments usually required in Italian opera; the orchestra of the Ancient Concerts had the reputation of being the best in London, but even this did not offer permanent employment to its members, and the standard of performance could not therefore compare with that of established continental orchestras of the time.

It was a period when foreigners generally regarded England as an unmusical country, but one nevertheless where fortunes were to be made. That we were devoid of good composers for the orchestra is true enough, and also that we showed little interest in the possibility of an English idea emerging. George Hogarth writing of Stephen Storace said: 'In writing for the orchestra, he has followed his Italian models in the simplicity of his score; but his accompaniments are full of grace and elegance, and want nothing but a slight infusion of German richness and variety to be everything that could be desired.' There was no recognition of English taste existing in its own right, only a shifting of favour from Italian taste towards that of the Germans. In the midst of this, established German players like Salomon still retained their artistic reputation, but the lack of novelties comparable with the Haydn symphonies of 1791 and 1794 influenced adversely the public response to orchestral music. So far did interest decline that by 1828 Fenimore Cooper saw 'respectable artists such as would be

FOUNDING OF PHILHARMONIC SOCIETY

gladly received in our orchestras walk the streets and play the music of Rossini, Mozart, Beethoven, Meyerbeer, Weber, etc., beneath your windows'.

This impoverishment of the rank and file is in accordance with the general trend of the *laisser-faire* policy of the age, and it would have been more marked — possibly even destroying all hope of competent orchestral playing — had not the professional musicians in London resolved in 1813 on a reform that aimed to place the art and their profession on a firm basis. This reform was the foundation of the Philharmonic Society of London, to which orchestral music in England owes a greater debt than to any other influence. The objects of the Society are set out in the preliminary announcement issued by the small group of musicians who formulated the idea, in which it is stated that:

> The want of encouragement, which has for many years past been experienced by that species of music which called forth the efforts, and displayed the genius of the greatest masters, and the almost utter neglect into which instrumental pieces in general have fallen, have long been sources of regret to the real amateur and to the well-educated professor: a regret which, though it has hitherto proved unavailing, has not extinguished the hope that persevering exertions may yet restore to the world those compositions which have excited so much delight, and rekindle in the public mind that taste for excellence in instrumental music which has so remained in a latent state. In order to effect this desirable purpose, several members of the musical profession have associated themselves, under the title of The Philharmonic Society, the object of which is to promote the performance, in the most perfect manner possible, of the best and most approved instrumental music, consisting of Full Pieces, Concertantes for not less than three principal instruments, Sextetts, Quintetts and Trios; excluding Concertos, Solos and Duets; and requiring that vocal music, when introduced, shall have full orchestral accompaniments, and shall be subjected to the same restrictions.

THE PHILHARMONIC PERIOD

The point of view of the promoters of the Society is sufficiently obvious from this announcement. They were professional instrumentalists desirous of recovering an almost lost market for their skill. They included Jean Baptist Cramer and Charles Neate, both of whom were pianists, the former a son of Wilhelm Cramer, who had come to London from Stamitz's famous orchestra at Mannheim, and the latter a friend of Beethoven. In answer to their appeal came the majority of reputable London instrumentalists, all willing to help in the direction of the Philharmonic Society or to play in its orchestra; several of the best players in London offering to play without a fee. The Society brought a new spirit into London concert life; hitherto orchestral concerts had been mainly commercial — the fact that much music of a high quality was introduced to the public by this means came from a demand among the aristocracy for the new types of symphony and concerto, which many of them heard on the continent when making the Grand Tour according to the system of the eighteenth century. This demand had fallen partly as a result of the Napoleonic War and partly owing to the reform of educational policy at Oxford. The Ancient Concerts were amateur controlled on a non-profit-making basis, but their adherence to the twenty years' rule operated against the interests of professional musicians, who wanted to give performances of their own music after the manner of the eighteenth century concert promoters. It was no longer profitable to compete with other performer-composers as the Professional Concert and Salomon had done; now it was a time for a sinking of individual interests in an endeavour to recover sufficient public interest in orchestral novelties to make these once more possible of performance. With this object the rules of the Philharmonic Society were formed; rules calculated to effect artistic results instead of pecuniary results.

> The Society to consist of thirty members and an unlimited number of Associates, from whom all future Members shall be chosen. Members and Associates to pay an annual subscription of three guineas.

FOUNDING OF PHILHARMONIC SOCIETY

> The Subscription to the Concerts, eight in number, to be four guineas, and for resident Members in the families of subscribers, two guineas each. No tickets to be transferable. Seven Directors to be annually chosen from among the Members for the management of the Concerts.
>
> No Member or Associate shall receive any emolument from the funds, all money received being appropriated only to the public purposes of the Society; nor shall any Member or Associate receive any pecuniary recompense for assisting at the Concerts.
>
> There shall not be any distinction of rank in the orchestra and therefore the station of every performer shall be absolutely determined by the leader of the night.

All those who enlisted under the banner of the new society were professional musicians; their names and qualifications — pianist, vocalist, violinist, composer, etc. — are given both by Hogarth (1862) and Myles Birket Foster (1912) in their histories of the Philharmonic Society of London. It was an adventure in many ways original, for previously amateur subscribers had guaranteed the success of new ventures in music; here was a case of professionals combining to reform their art, and in the process, not only agreeing to forgo their fees, but to eliminate distinctions of rank or professional standing in the cause of orchestral unanimity. The instrumental soloist, however, was becoming more important, and had the Philharmonic Society kept to the original rule of playing concerted movements only, and excluding solos, the result would have been disastrous, but professional sagacity overrid the rules, concertos began to be played at Philharmonic concerts in 1819, and the artistic progress of the Society during the first half of the nineteenth century was in harmony with the general development of the art of orchestral music abroad, and led the way in this country.

One reason why the glorification of a single individual was unacceptable to the founders of the Philharmonic Society lies in the fact that such a system was contrary to the current prac-

THE PHILHARMONIC PERIOD

tice of orchestral playing in England. Here the method of orchestral control was still by means of a leading violinist and a colleague who 'presided at the pianoforte'; London musicians were strongly opposed to a single individual controlling their efforts by beating time with a baton.[1] The orchestral player had not yet come to regard himself as separate from the chamber-music player. Chamber music was distinguished at the beginning of the nineteenth century by its complete freedom from the tyranny of the figured bass, and formed therefore a much better medium of expression than the orchestra could ever be under the system of dual control. Concert halls, moreover, were not yet so large that string quartets, written for performance in the private apartments of eighteenth-century patrons, would lose their appeal under the necessity of being heard at the far end of the halls. This musical problem developed later.

Under the system of dual control the pianist still had his place in the orchestra. By the nature of things it was a leading place, and although the pianoforte was self-sufficient in the home, and already a source of profit to enterprising musicians who were devoting their attentions to the exploitation of this new market, it was not regarded as self-sufficient in the concert-hall until well into the century. Eminent pianists like Muzio Clementi and J. B. Cramer, whom we know as the authors of excellent studies in pianoforte technique and as pioneers in composition for the pianoforte in its own right, nevertheless had an interest in the orchestra for public performances. Clementi and Cramer were both original members of the Philharmonic Society, and Clementi presided at the pianoforte at the first concert of the Society, held on Monday, March 8th, 1813. The leading violinist was Salomon, veteran of so many triumphs in London subscription concerts, and the programme was made up of equal parts of what we would now call orchestral music and chamber music, with a couple of concerted vocal pieces introduced to provide relief from the general mass of purely

[1] It was not unknown, however. See page 67 and the illustration facing page 65.

FOUNDING OF PHILHARMONIC SOCIETY

instrumental tone. The instrumental pieces in the chamber music idiom were a string quartet by Mozart, a string quartet by Boccherini, and a serenade for six wind instruments by Mozart. The particular works cannot be identified further from the printed programme. The same is true of the two symphonies contained in this programme: one was by Beethoven and the other by Haydn, but no number, key, or statement of the nature of their movements is given. The remaining orchestral pieces were Cherubini's overture to *Anacreon* and Haydn's *Chaconne, Jomelle and March*.

For years, until the coming of Spohr and Mendelssohn, such programmes were typical of the Philharmonic Society. Haydn, Mozart, Gluck, Cherubini, Beethoven, Clementi, and A. Romberg were the composers whose names most commonly appeared on their programmes. The common belief that Beethoven was a neglected composer is not borne out by a perusal of the Philharmonic Society's programmes, for in the first five years of that Society's activities, embracing forty concerts, his name appears forty-nine times against thirty-four of Cherubini's. Cherubini's popularity in England has in fact been exaggerated, for not only is he outnumbered in programme appearances, but the works performed are of less substance than Beethoven's. Beethoven's symphonies, overtures, and chamber-music compositions appear on every programme of the Society from 1813 to 1818, excepting only four, and he is outnumbered only by Haydn, with fifty-two performances to his credit, and Mozart, with the amazing number of one hundred. Handel's name appears once only, and that is for a vocal scena sung by Braham; but that is no argument against Handel's popularity, for the Philharmonic Society came into existence as a counterblast to the Concert of Antient Music, which relied greatly on the popularity of Handel's music for its continued success.

This was no new state of affairs in London musical circles. Ever since 1776 the Ancient Concerts had been the most influential musical gatherings in London, enjoying the support

THE PHILHARMONIC PERIOD

of Royalty and the cream of society, but restricted by their aims to performances of music that had stood the test of time. Their policy was conservative but sound. Against them it had always been necessary for the modern composer to set up his own organization if his works were to be heard at all. The Philharmonic Society was the final expression of this necessity, and a true successor of the subscription concerts of such eminent eighteenth-century organizations as Bach and Abel's concerts, the Professional Concert, and Salomon's concerts. The policy was the same, for just as in the eighteenth century it is possible to trace the works performed, when not by members of those organizations, to personal friends of theirs and their aristocratic patrons, so it is possible to trace personal friendships between members of the Philharmonic Society and all the composers whose works appeared on their programmes. Mozart was known to several of them, and especially to Thomas Attwood, his friend and pupil. The glory of Haydn's London triumphs of 1791 and 1794 still shone on Salomon; Viotti had been associated with Cherubini in the direction of the Theatre de Monsieur in Paris from 1789 to 1791, and Charles Neate knew Beethoven intimately and claimed to be his only English pupil. The atmosphere of the Philharmonic Society was that of a professional club; they knew all the trends of music in their day and were anxious to make them known to the wider public.

The original orchestra of the Philharmonic Society was more up to date in construction and of finer quality than that of its great rival the Concert of Antient Music. The latter comprised in addition to strings, 4 oboes, 4 bassoons, 4 horns, 2 trumpets, a trombone, and drums. From this we may assume that their ideas of tone-balance remained much as they were in Handel's time. The orchestra used at the Philharmonic Society's fourth concert in 1813 is given by Foster as Salomon (leader), Spagnoletti, F. Cramer, and Moralt; violas, Mountain, W. Griesbach, and Sherrington; 'cellos, Ashley, Crouch, and Robert Lindley; double bass, Henry Hill, senr.; flute, Ashe; oboes, F. Griesbach and M. Sharp; clarinets, Mahon, Oliver, and Kramer; bas-

FOUNDING OF PHILHARMONIC SOCIETY

soons, Holmes and Tully; horns, Joseph and Peter Petrides; J. B. Cramer presided at the pianoforte. In addition to these players solo parts in the chamber-music pieces were played on the piano by Charles Neate and Ludwig Berger; G. A. P. Bridgetower, the mulatto violinist, appeared as first violin in a Beethoven string quintet, and H. Gattie as second viola. Every man therefore was an efficient player; and three of these violinists, Salomon, Cramer, and Spagnoletti, acted as leaders at numerous concerts. By 1860 the Philharmonic Society's orchestra had grown to sixty-six players, made up of 12 1st violins, 12 2nd violins, 8 violas, 8 'cellos, 8 basses, 2 each of flutes, oboes, clarinets, and bassoons, 4 horns, 2 trumpets, 3 trombones, and 1 drummer[1] — the normal classical orchestra of the nineteenth century. By this time the work of Berlioz and Wagner had made an orchestra even of those dimensions appear small, but the establishment of such an orchestra had involved a hard struggle; a struggle, however, that was borne fearlessly and with much honour, especially in its early stages. A deciding factor in the establishment of this time-honoured orchestra was the personality of its early members: they were men with high ideals, endeavouring to improve their art and the dignity of their profession at a time when social changes greater than they imagined were about to take place. Under the circumstances they were sure to make some mistakes, but the history of their orchestra is nevertheless the most important single factor in the advancement of musical taste in England from 1813 until the renaissance of British music near the end of the nineteenth century.

[1] Myles Birket Foster, *The History of the Philharmonic Society of 'London, 1813-1912.*

TEMPORA ET MORES

IT needed more than musical ability to be successful in London concert circles in the early nineteenth century; it needed independence and business enterprise. Salomon was successful because he had all these qualities; lacking any one of them, he would have sunk into the anonymous rank and file of orchestral players, earning a precarious livelihood in some theatre orchestra with occasional extra pickings at balls, subscription concerts or Lenten oratorios. He lived through a most difficult transitional period, but remained master of the situation, and by his enterprise helped Haydn to achieve his highest flights in symphonic composition. Had Mozart disregarded his father's advice and visited London when the Storaces invited him to do so, his contribution, too, might have been greater.

But Mozart could not have been expected to understand the way in which music had developed in England. London musicians had passed beyond the single business of concert promotion. Some musicians not only composed and performed music but printed and published it; they even manufactured instruments. The borderline between music as a profession and music as a business had in our most successful men ceased to exist. Clementi made three fortunes as a pianoforte manufacturer and music publisher, and his pupil J. B. Cramer also founded a successful music publishing house; they were nevertheless eminent as composers and players, and indeed relied on their musical reputations to attract customers to their trading establishments. It was one effect of the social revolution coming hard on the heels of the Industrial Revolution in this country. The wealthy middle classes offered better prospects of a livelihood to musicians than aristocratic support, and it was possible for musicians to win recognition of social equality with the middle classes by the exercise of industry and a good head for business.

TEMPORA ET MORES

Clementi came to England as a boy and was privately trained by a benefactor named Peter Beckford.[1] He spent his youth in the seclusion of this gentleman's Dorset residence until at the age of twenty-one he came to London, a pianist conspicuously superior to any other in the capital. His education had been different from that of other musicians, through being concentrated on the pianoforte as a solo chamber instrument and not as a centre round which to group an orchestra. London forced him to play in orchestras, however, for there was at that time no public demand for anything like a modern pianoforte recital. Clementi conducted the Italian Opera in London from 1777 to 1780 and then set out on a continental tour that established him as one of the world's foremost executants on the pianoforte. His style of performance and of composition were pianistic — not reminiscent of the old harpsichord — and it is unfortunate that Mozart's condemnation of his style ('a mere mechanician, strong in runs of thirds, but without a pennyworth of feeling or taste') should be so often accepted as true. The greatest benefit a composer could bestow on the new instrument at that time was to crystallize the general features of the new sonata form and adapt it to the distinctively new instrument. This Clementi did, and it is sufficient for his claim to fame that he did it. The work of the romantic pianoforte sonata composers, starting with Beethoven, was rendered easier by Clementi's standardization of the form that they were so soon to mould into a flexible medium for their own original ideas. Besides this, there was a need for a system of teaching the pianoforte, without which its distinctive features could not be exploited by the amateur. Clementi did all this, and the art of pianoforte playing was well established at the time of his death. His pupil J. B. Cramer carried on the work his master had begun.

Yet Clementi lacked something essential to a great artist. The key to it is to be found in his treatment of John Field, in whose exquisite touch on the keyboard Clementi saw not the future of romantic pianism but only a means of selling his

[1] A cousin of the author of *Vathek*.

pianos. He kept young Field, the wonder of his age, in his shop to display the quality of his instruments to customers, took him abroad for the same purpose, and finally left him, although in delicate health, at his St. Petersburg agency. Such behaviour is not that of an artist but of a business man, and no doubt this quality of hard-boiled commercialism shown by the most successful of the London musicians encouraged the belief abroad that England was an unmusical country. The popularity his own skill as a pianist brought him Clementi used to attract purchasers of his pianos and his music, and to those on the continent that was a degradation of his art.[1] But such a view is fatuous; the musician must live, and Clementi's organization of all the resources surrounding the new pianoforte was in the end more fruitful than employment in an opera house could have been. English pianos at this period were good but expensive; music copies, too, were expensive in this country, but better printed than abroad. Teachers of the pianoforte were well paid by wealthy middle-class parents who wished to improve the social attraction of their daughters. Clementi and Cramer established the nineteenth-century music teacher securely in his profession, and by so doing influenced the standard of amateur performance in the Victorian home. This had its effect on the taste of Victorian audiences. Musicians also, as business men, became more directly concerned with the ordinary responsibilities of citizenship than foreign musicians previously living in London had been.

How important this had become is evident from the position of Viotti. Of all the violinists of his time, G. B. Viotti is generally recognized as the finest. He did for the violin concerto what Clementi did for the pianoforte sonata, and his knowledge of the resources of his instrument has led him to be regarded as the father of modern violin playing. Yet he occupied a position subordinate to Salomon at the first Philharmonic concert in 1813. His work nevertheless was not unrecognized by his colleagues, for in the first three years of the

[1] But not unknown on the Continent.

TEMPORA ET MORES

society's existence four of Viotti's compositions were performed at their concerts, and on each of these occasions Viotti acted as leader. But after 1815 Viotti's name does not appear on their programmes, although he continued his musical career until his death in 1824, living part of the time in Paris (where he was greatly admired) and partly in London.

Viotti, in fact, presents a problem that was to become common in the nineteenth century: the problem of a musician of merit left to his own resources and unable to cope with the economic difficulties of his period. Viotti had no understanding of current affairs; he was the only musician to be driven out of Paris by the Revolution, and in England he failed to adapt himself to the business of an independent musician like Salomon; nor could he, like Clementi, successfully engage in trade, for his efforts to augment his income in England by means of a wine business resulted in the accumulation of debts. He was temperamentally incapable of adjusting his ideas to the facts of his times; yet he was an honest man, amiable, and greatly respected.

The secret of his failure in everyday life is revealed in his compositions. There is a fine nobility in his music that, even though it lacks striking originality, reveals a thinker striving to express himself rather than gain the ear of the multitude. Constantly his desire for seclusion came in conflict with the necessity of keeping his name before the public. In Paris he enjoyed the favour of the best violinists, but had to be drawn into his rare public appearances; in London, as we have seen, he led the Philharmonic orchestra only when he had a new composition to produce, and (most significant of all) when by some whim of British bureaucracy he was exiled for three years from this country as *persona non grata* to the state, he employed his time abroad not in a concert tour to the various capitals of Europe, but in the seclusion of a small German village where he devoted himself to composition. It was to become an urgent problem for musicians during the nineteenth century to know how to order their lives so as to carry out their art work; finally they split into three main groups — the solo

THE PHILHARMONIC PERIOD

performers, who travelled from place to place appearing in the great cities; the academic musicians, who settled in the university cities and such as had established musical conservatoires, and the composers, who needed long periods of seclusion in order to concentrate on their creative work, but who nevertheless found it expedient to travel occasionally, and to make public appearances as performers. Viotti stood at the parting of these ways, unable to choose a path that would lead him both to security of living and the fulfilment of his artistic desires. In comparison with Beethoven, Berlioz, or Wagner, Viotti's problem tends to recede into the background, but it was the same problem.

These, however, were the affairs of artists of the first rank, and not of the group of capable players who formed the main body of the Philharmonic Society of London at its inception in 1813. The French Revolution had led to the Napoleonic Wars and Europe was consequently still unsettled; yet all over Europe music continued to be performed unmolested. The economic difficulties which beset musicians were no worse than those that beset all other professional men — high taxation, monetary inflation, and such-like evidences of rash national policy. Even in England paper money began to appear, and inflation aggravated still further the distress of the industrial workers. But orchestral music was for the well-to-do only. The subscription to the Philharmonic Society was three guineas a year for members and associates, and for the concerts — eight in number — four guineas. (This price of admission to London concerts had remained steady at half-a-guinea since Handel's time.) At first the orchestra was led entirely by foreigners — Salomon, Franz Cramer, Viotti, Spagnoletti, and Vaccari among the violinists, and the pianoforte played alternatively by Clementi and J. B. Cramer. Not until 1816 does an English name appear on the programmes at the piano, and then it is that of Attwood, Mozart's favourite pupil. Sir George Smart followed, and gradually other English names begin to appear — Dr. Crotch and Griffin at the piano, and J. D. Loder as

TEMPORA ET MORES

violinist at the sixth concert of the 1817 season mark the break with the German, Italian, and French monopoly of leadership. But it must not be assumed that there was any awakening of a national rivalry for control of our English destiny in orchestral music, for although British musicians abroad praised London musical life constantly — Attwood, Linley, and Storace to Mozart, and Neate to Beethoven — for a long time it was taken for granted in England that imitation of foreign examples was the only way to success in orchestral composition. George Hogarth, as late as 1835, could write in his book on *Musical History, Biography and Criticism*:

> The composition of orchestral music either for a full orchestra, or in the form of concerted pieces for instruments, has not yet been successfully cultivated in England. We have no symphonies, quartets, or quintets, that have attracted attention even among ourselves; and our dramatic composers, though some of them are able to employ the orchestra effectively as an accompaniment, hardly ever fail, in their overtures, to show their deficiency in instrumental composition. To excel in this branch of the art demands a depth and variety of knowledge and a command of the resources of harmony which, till very lately at least, has been unattainable by the imperfect means of education which England has furnished to musical students; and they have, moreover, to contend with the stupendous works of the German school, the excellence of which appears to the public, as well as to themselves, to be unapproachable. This produces the double disadvantage of depressing their energies, and of preventing their productions from having an indulgent, or perhaps even a fair, hearing. If a symphony, an overture, or a quartet, by a native aspirant for musical honours is performed in public, the question ought to be, not whether it is comparable with the work of Haydn, Mozart, or Beethoven, but whether it contains sufficient originality, ingenuity, learning, and beauty, to please in spite of the defects incident to youth and inexperience, and to give good assurance of future excellence. There is even more pleasure in listening to music of this description, and

THE PHILHARMONIC PERIOD

in contributing by liberal approbation, to the encouragement of rising talent, than of enjoying the most consummate work of the greatest master. The most severe and captious criticism proceeding from inferior artists and superficial amateurs; while the most eminent and accomplished musicians show alacrity in discovering beauties, and in putting the kindest construction on the existence of faults. Our musical students have now the means to ample instructions;[1] and its fruits are apparent in recent instrumental compositions of such merit as to show that their authors, if encouraged to persevere, are capable of raising the English school to distinction even in this arduous department of the art.

The compositions of such men as Cipriani Potter, Griffin, and J. B. Cramer have not justified any hopes that Hogarth may have entertained for their future — it is as directors of the orchestra that they are to be remembered. Certainly the Philharmonic Society of London remained true to its ideal of performing music on account of its artistic merit and not for the profit of its members. It may be for this reason that the Philharmonic Society declined to combine with the Professional Concert in 1815 — both organizations were engaged in similar work, but the Professional Concert was a concern run for private profit and the Philharmonic Society was not. Changes came about, but slowly, for there was a very strong determination not to deviate from the rules laid down when the Society was formed. Some of the rules had to be modified, however, as the new romantic spirit in music began to establish itself. The rules forbade solo songs with pianoforte accompaniment, but Mozart's *Dove Sono* came in 1816 and Beethoven's *Adelaide* found a place in the programme for 1817, with Sir George Smart as accompanist. Some attempt was made to compromise with the rules by calling *Adelaide* a cantata, but it is so obviously a song that the audience cannot have had any doubts about the change of policy implied by its inclusion in the programme. In

[1] A reference to the Royal Academy of Music founded in 1822.

TEMPORA ET MORES

1819 the first concertos appear on the Philharmonic programmes. *A Fantasia for Clarinet and Orchestra* by Baermann at the second concert of that year, a Mozart Pianoforte Concerto at the fifth concert, and a first performance of a Pianoforte Concerto by J. B. Cramer at the sixth concert. The rule 'excluding Concertos, Solos and Duets, and requiring that vocal music, when introduced, shall have full orchestral accompaniments and shall be subjected to the same restrictions' had, by 1819, been dropped.

The greatest reform of all, however, had to wait until 1820, when Louis Spohr came to London. At the first concert of that year he played his violin concerto, *Nello Stilo Drammatico*, at their second and sixth concert he led the string quartet in his chamber compositions, and at the third concert of that year he conducted the first performance of one of his own symphonies. He was given every facility, therefore, to present his own compositions to the London public in accordance with his own ideas, even to the extent of conducting his symphony with a baton. Spohr had his way, but not without considerable trepidation, as we can tell from an entry in his *Autobiography*:

> I resolved, when my turn came to direct, to attempt to remedy this defective system! At the morning rehearsal on the day I was to conduct (Monday, June 19), I took my stand with a Score at a separate music-desk in front of the orchestra, drew my directing baton from my coat pocket, and gave the signal to begin. Quite alarmed at such novel procedure some of the Directors would have protested against it; but when I besought them to grant me at least one trial, they became pacified. The triumph of the baton, as a time-giver, was decisive, and no one was seen again seated at the pianoforte during the performance of Symphonies and Overtures.

From that time onwards the words 'At the pianoforte Mr. — ' do not appear on the programme of the Philharmonic Society, but the name of the leader appears together with that of the conductor.

ENGLISH RELATIONS WITH BEETHOVEN

No more creditable record of steadfast adherence to an artistic ideal can be found in the early nineteenth century than the Philharmonic Society's policy of spreading the music of Beethoven. They kept his compositions always before their public, commissioned works from him, invited him to London, and, finally, responded generously to his appeal for financial assistance during his last illness; they found him difficult in business negotiations, liable to misinterpret their good intentions, a man facile with his promises, but slow in their performance; a man whom, had he been one of themselves, they would have avoided, but whom they nevertheless supported because of his high artistic ideals, and because, even when he behaved most badly towards them, he did so with good intentions.

The more we look into the aspirations of Beethoven the more we are astounded at his achievements. He not only set himself to produce music of an outstanding quality, but succeeded in evolving a system of marketing his genius which left him with more freedom of action than any of his immediate forerunners had enjoyed. He reversed the usual custom of society so far that instead of fawning upon the moneyed classes he actually demanded that they should acquiesce in his moods and supply his needs. The evolution of such a system of living was in itself a revolution, and we may well ask by what means he brought it about.

It is significant that when Beethoven appealed to the aristocracy of Vienna or to the directors of the Court Theatre — who also were drawn exclusively from the aristocracy — he was able to sound a loftier theme than that of his own personal needs. He could appeal to their patriotism. They supported him because they wished to see the art of the German-speaking peoples rise triumphant over all others. The French were

ENGLISH RELATIONS WITH BEETHOVEN

politically in the ascendant: the campaigns of Napoleon had shaken Europe to its foundations, but the Germanic peoples kept alive their national pride by maintaining their distinctive culture. French music they had never admired, and Italian music they began to discard even before the threat of foreign domination by force of arms. Beethoven, writing to the directors of the Court Theatre, Vienna, was able to approach them in a dignified manner, representing himself as one with whom it would be an honour for a national theatre to be associated.

> The mere wish to gain a livelihood has never been the leading clue that has hitherto guided the undersigned on his path. His great aim has been the interest of art and the ennobling of taste, while his genius, soaring to a higher ideal and greater perfection, frequently compelled him to sacrifice his talents and profits to the Muse. Still, works of this kind won for him a reputation in distant lands, securing him the most favourable reception in various places of distinction, and a position befitting his talents and acquirements.
>
> The undersigned does not, however, hesitate to say that this city [Vienna] is above all others the most precious and desirable in his eyes, owing to the number of years he has lived here, the favour and approval he has enjoyed from both high and low, and his wish fully to realise the expectations he has had the good fortune to excite, but most of all, from his patriotism as a German. Before, therefore, making up his mind to leave a place so dear to him, he begs to refer to a hint which the reigning Prince Lichnowsky was so kind as to give him, to the effect that the directors of the theatre were disposed to engage the undersigned on reasonable conditions in the service of the theatre, and to ensure his remaining in Vienna, by securing to him a permanent position, more propitious to the further exercise of his talents etc.

The remainder of the letter is devoted to a description of the work Beethoven would be prepared to do, and the remunera-

THE PHILHARMONIC PERIOD

tion he would expect to receive. Beethoven did not get the post he sought, but his letter did not bring down the wrath of Viennese nobility upon him in the way that a similar letter written, say, by Haydn or Mozart, would have done. Let us, when we think of Beethoven's conceits, his overestimation of his own importance, and his unjust treatment of friends who tried to advance his cause, remember that he achieved the extraordinary task of commanding respect for one of his profession from many who regarded nobility as an accident of birth rather than a quality of mind.

But Beethoven's letter to the directors of the Court Theatre was written in 1807. By 1813, when the Philharmonic Society of London was inaugurated, Beethoven was growing desperate. His illness was wearing down his temper, and he had the added responsibility and expense of his nephew Carl. Beethoven lived more and more on his nerves as practical difficulties increased, and his art — deepening as his genius came more fully towards maturity — made increasing demands on his powers of concentration, drained him of his mental stamina, and necessitated periods of seclusion and rest in the country, where in due course the germs of succeeding masterpieces could formulate in his mind. The evidence is in his sketch-books. All the time this artistic productive process was going on Beethoven was plagued by the necessity of obtaining some financial benefit from the compositions that emerged, and in this he played a lone hand, constantly involved in the complications of publishers' rights and his own expectations, the dangers of piracy, and misunderstandings with the various friends through whom he approached interested parties abroad.

Because of the distance between London and Vienna, the length of time taken by the post in those days and the necessity of trusting someone to act as his agent, Beethoven's dealings with the Philharmonic Society of London were more than usually troublesome. They were made more difficult by reason of Beethoven's habit of accusing Neate and Cramer of working against his interests, when, in fact, they were mainly instrumental in procuring for him a contract for the payment of seventy-five guineas in return for the right of first performance of three overtures in 1815 — *King Stephen, Ruins of Athens,* and the *Overture in C, Op.* 115. Such announcements as the following, which appeared in 1814, must also have been disturbing:

> Herr Maelzel, now in London, on his way thither performed my *Battle Symphony* and *Wellington's Battle of Vittoria* in Munich, and no doubt he intends to produce them at London concerts, as he wished to do in Frankfort. This induces me to declare that I never in any way made over or transferred the said works to Herr Maelzel; that no one possesses a copy of them, and that the only one verified by me I sent to His Royal Highness the Prince Regent of England. The performance of these works, therefore, by Herr Maelzel is either an imposition on the public, as the above declaration proves that he does not possess them, or if he does, he has been guilty of a breach of faith towards me, inasmuch as he must have got them in a surreptitious manner.
>
> But even in the latter case the public will still be deluded, for the works which Herr Maelzel performs under the titles of *Wellington's Battle of Vittoria* and *Battle Symphony* are beyond all doubt spurious and mutilated as he never had any portion of either of these works of mine, except some of the parts for a few days.
>
> This suspicion becomes a certainty from the testimony of various artists here, whose names I am authorized to give if necessary. These gentlemen state that Herr Maelzel, before he left Vienna, declared that he was in possession of these works, and showed various portions, which, however, as I have already proved, must be counterfeit. The question whether Herr Maelzel be capable of doing me such an injury, is best solved by the following fact. In the public papers he named himself as sole giver of the concert on behalf of our wounded soldiers, whereas my works alone were performed there, and yet he made no allusion whatsoever to me.
>
> I therefore appeal to London musicians not to permit such

THE PHILHARMONIC PERIOD

a grievous wrong to be done to their fellow-artist by Herr Maelzel's performance of the *Battle of Vittoria* and the *Battle Symphony*, and also to prevent the London public being so shamefully imposed upon.

The law of libel was in those days as ineffective as the law of copyright. If Beethoven suffered from one, he could always have recourse to the other. Unfortunately his libels were often directed at the reputations of true friends. Charles Neate, as letters of his long preserved in the Berlin State Library testify, was a most faithful worker in Beethoven's cause, yet while he was endeavouring to promote performances of Beethoven's compositions in London, and arrange for their publication, Beethoven was decrying him in letters to Ries, who was at the same time in London. Yet it must not be assumed that Beethoven was activated by unworthy motives: he was extremely worried. His compositions were as much to him as a baby to a young mother, and the world was too big a place for them to be turned loose without some protection. So Beethoven grew impatient at any delay, and blamed himself equally with those to whom he had entrusted his compositions. 'I am almost ashamed to allude to the other works I entrusted to him [Neate], and equally so of myself, for having given them to him so confidingly, devoid of all conditions save those suggested by his own friendship and zeal for my interests.' So it was always with Beethoven: trusting almost recklessly at one time, and at another beset with harrowing doubts because the trusted one was not present to reassure him. Yet this must be said for Neate, that he continued to believe in Beethoven though the contents of Beethoven's letters ultimately became known to him.

It is possible that Ries fed Beethoven's suspicions concerning Neate, as he undoubtedly did about J. B. Cramer, for in a letter to Salomon in 1815 Beethoven said:

> I hear, indeed, that Cramer is also a publisher, but my scholar, Ries, lately wrote to me that Cramer not long since publicly expressed his disapproval of my works, I trust from no motive but that of being of service to art, and if

so I have no right to object to his doing this. If, however, Cramer should wish to possess any of my pernicious works, I shall be as well satisfied with him as with any other publisher; but I reserve the right to give these works to be published here, so that they may appear at the same moment in London and Vienna.

This last sentence explains the source of many of Beethoven's publishing difficulties. If a work was published anywhere it was only a matter of time before it was copied by some other printer and piratically reprinted. The only way to avoid this was by having the work simultaneously published in all countries where there would be a demand for it. Beethoven's correspondence shows that these arrangements were difficult to make: he had to rely on the honesty of his correspondents, some of whom he did not personally know, and he himself was unduly sensitive about his honour when the arrangements went wrong and a work appeared in one country before the publisher had got the same work ready for issue in another. Continuing the letter to Salomon partly quoted above, Beethoven adds:

> Perhaps you may also be able to point out to me in what way I can recover from the Prince Regent the expenses of transcribing the *Battle Symphony* on Wellington's victory at Vittoria to be dedicated to him, for I have long ago given up all hope of receiving anything from that quarter. I have not even been deemed worthy of an answer, whether I am to be authorised to dedicate the work to the Prince Regent: and when at last I propose to publish it here, I am informed that it has already appeared in London. What a fatality for an author! While the English and German papers are filled with accounts of the success of the work, as performed at Drury Lane, and that theatre drawing great receipts from it, the author has not one friendly line to show, not even payment for the cost of copying the work, and is thus deprived of all profit. For if it be true that the pianoforte arrangement is soon to be published by a German publisher, copied from the London one, then I lose both my fame and my honorarium. The well known

THE PHILHARMONIC PERIOD

generosity of your character leads me to hope that you will take some interest in the matter, and actively exert yourself on my behalf.

This wretched potboiler gave Beethoven more anxiety than any of his more characteristic works, and it is to the credit of the directors of the Philharmonic Society that they never performed it. From 1814 to 1823 Beethoven was writing to one or another of them asking if they could influence the Prince Regent towards the bestowal of some recognition for the dedication. Beethoven relied on the vanity of his aristocratic dedicatees to increase his profits on a work, but again it was a personal matter, depending on the character of the dedicatee. Had Beethoven made some inquiries about the reputation of the Prince Regent, he would probably not have sent the *Battle Symphony* to him, for Haydn had had difficulty in getting payment from him for his private performances when in London, until finally he sent a bill to Parliament, which paid immediately without question. The Prince's idiosyncrasies were well known, as indeed were those of the whole Royal Family: 'They are the damnedest millstones about the neck of any Government that can be imagined', said the Duke of Wellington. Beethoven succeeded in getting payment of forty ducats for the *Battle Symphony* from the London publisher, Robert Birchall, but his hopes of a donation from the Prince Regent, even after he became King George IV, came to nothing. Beethoven kept referring to the Prince in his letters to all associated with the Philharmonic Society because this Society enjoyed the Prince's support. There is a letter from Beethoven to Ries, dealing with the Prince Regent, but quoted here because it reveals particulars of Beethoven's financial position in 1816:

> My answer has been too long delayed: but I was ill, and had a great press of business. Not a single farthing is yet come of the ten gold ducats, and I now almost begin to think that the English are only liberal when in foreign

countries. It is the same with the Prince Regent, who has not even sent me the cost of copying my *Battle Symphony*, or one verbal or written expression of thanks. My whole income consists of 3,400 florins in paper-money. I pay 1,100 for house rent, and 900 to my servant and his wife, so you may reckon for yourself what remains. Besides this, the entire maintenance of my young nephew devolves on me. At present he is at school, which costs 1,100 florins, and is by no means a good one, so that I must arrange a proper household and have him with me. How much money must be made to live at all here! And yet there seems no end to it — because! — because! — because! — but you know well what I mean.

Some commissions from the Philharmonic would be very acceptable to me, besides the concert.

But when the Society's response to this appeal came to hand in the following year, 1817, Beethoven tried to drive a difficult bargain and their negotiations broke down.

The Society had for some time had in mind an intention to invite Beethoven to London, and Beethoven had often expressed a wish to visit them. They therefore made him an offer of 300 guineas to come to London and direct two symphonies to be composed by him for the Society. Two years previously Cherubini had accepted a similar contract to supply and direct a symphony, an overture and a vocal trio, *Et Incarnatus est*, for a fee of £200. Beethoven therefore was treated equally generously by the Society in their 1817 offer. But Beethoven wanted more. His letter to Ries of July 9th runs:

My dear Friend,

The proposals in your esteemed letter of the 9th of June are very flattering, and my reply will show you how much I value them. Were it not for my unhappy infirmities which entail both attendance and expense, particularly on a journey to a foreign country, I would unconditionally accept the offer of the Philharmonic Society. But place yourself in my position, and consider how many more obstacles I have to contend with than any other artist, and then judge whether my demands (which I annex) are unreasonable. I beg you will convey my conditions to the Directors of the above Society, viz.:

1. I shall be in London early in January.
2. The two grand new Symphonies shall be ready by that time; to become the exclusive property of the Society.
3. The Society to give me in return 300 guineas and 100 for my travelling expenses, which will, however, amount to much more, as I am obliged to bring a companion.
4. As I am now beginning to work at these grand Symphonies for the Society, I shall expect that (on receiving my consent) they will remit me here the sum of 150 guineas, so that I may provide a carriage and make my other preparations for the journey.
5. The conditions as to my non-appearance in any other public orchestra, my not directing, and the preference always to be given to the Society on an offer of equal terms by them, are accepted by me; indeed, they would at all events have been dictated by my own sense of honour.
6. I shall expect the aid of the Society in arranging one or more benefit concerts in my behalf, as the case may be. The very friendly feeling of some of the Directors in your valuable body, and the kind reception of my works by all the artists, is a sufficient guarantee on this point, and will be a still further inducement to me to endeavour not to disappoint their expectations.
7. I request that I may receive the assent to and confirmation of these terms, signed by three Directors in the name of the Society. You may easily imagine how much I rejoice at the thoughts of becoming acquainted with the worthy Sir George Smart and seeing you and Mr. Neate again: would that I could fly to you instead of this letter.

Your sincere well wisher and friend
LUDWIG VAN BEETHOVEN.

If Beethoven did not entirely trust the directors of the Philharmonic Society, it is possible that they too did not entirely trust him. They would not agree to the suggested increase and advancement of cash, but repeated their previous offer; to this Beethoven would not agree, and he never visited London; but he was long toying with the idea, even though he had declined the Society's offer, characteristically chanting his goodwill, bemoaning his health (which really was considerably worse) and suspecting Neate and Cramer. In 1818 he wrote to Ries:

> Pray request Neate in my name, to make no public use of the various works of mine that he has in his hands, at least not until I come. Whatever he may have to say for himself, I have cause to complain of him.
> Potter called on me several times; he seems to be a worthy man and to have a talent for composition. My wish and hope for you is that your circumstances will daily improve. I cannot alas! say that such is the case with my own ... I cannot bear to see others want, I must give; you may therefore believe what a loser I am by this affair. I do beg that you will write to me soon. If possible, I shall try to get away from this earlier, in the hope of escaping utter ruin, in which case I shall arrive in London by the Winter at latest. I know that you will assist an unfortunate friend. If it had only been in my power, and I had not been chained to this place, as I always have been by circumstances, I certainly would have done far more for you.
> Farewell; remember me to Neate, Smart and Cramer. Although I hear that the latter is a counter subject, both to you and to myself, still I rather understand how to manage people of that kind, so notwithstanding all this we shall yet succeed in producing an agreeable harmony in London. I embrace you from my heart.
>
> Your friend,
BEETHOVEN.

Again the proposal to visit London came to nothing, although the Philharmonic Society continued to perform his works regularly. The slightest criticism, however, had an adverse effect on Beethoven, and it would appear from a letter to Ries

THE PHILHARMONIC PERIOD

in 1819 that the unfortunate Neate had again come under suspicion for his honesty:

> I expect every day the text of a new Oratorio which I am to write for our Musical Society here and no doubt it will be of use to us in London also. Do what you can on my behalf, for I greatly need it. I should have been glad to receive any commission from the Philharmonic but Neate's report of the all but failure of the three overtures vexed me much. Each in its own style not only pleased here, but those in E flat major and C major made a profound impression, so that the fate of those works at the Philharmonic is quite incomprehensible to me.

He was not satisfied, either, with the amount offered for the *Ninth Symphony*, which was the subject of his next business deal with the Philharmonic Society. The *Ninth Symphony* had been germinating in his mind since 1815; it began to take shape more firmly after 1818, and by 1822 it was known to Ries that this latest symphony was approaching completion. In April he received a letter from Beethoven which said:

> My dearest and best Ries,
> Having been again in bad health during the last ten months I have hitherto been unable to answer your letter. I duly received the £26 sterling, and thank you sincerely. I have not, however, yet got the Sonata you dedicated to me. My greatest work is a Grand Mass that I have recently written. As time presses, I can only say what is most urgent. What would the Philharmonic give me for a Symphony?
> I still cherish the hope of going to London next spring, if my health admits of it! Etc.

At a Philharmonic Directors' Meeting on November 10th, 1822, it was resolved to offer Beethoven the sum of £50 for a MS. Symphony, to be delivered in March 1823; all rights were to revert to the composer eighteen months after the date of receiving the work. This resolution was duly reported to Beethoven by Ries, and accepted by the composer in a letter dated December 20th, 1822:

ENGLISH RELATIONS WITH BEETHOVEN

My dear Ries,
I have been so over-burdened with work, that I am only now able to reply to your letter of November 15th. I accept with pleasure the proposal to write a new Symphony for the Philharmonic Society. Although the price given by the English cannot be compared with those paid by other nations, still I would gladly write even gratis for those whom I consider the first artists in Europe — were I not still, as ever, the poor Beethoven.
If I were only in London, what would I not write for the Philharmonic! For Beethoven, thank God! can write — if he can do nothing in the world besides! If Providence only vouchsafes to restore my health, which is at least improving, I shall then be able to respond to the many proposals from all parts of Europe and even North America, and thus perhaps be some day in clover.

The money agreed on was sent without delay, but the symphony was longer in completion than had been anticipated; partly this was due to the enormous proportions of the symphony itself and partly to Beethoven's health. 'Do not be uneasy', he wrote to Ries the following April, 'you shall shortly receive the Symphony; really and truly, my distressing condition is alone to blame for the delay.' Beethoven suffered in addition to his other troubles from a distressing inflammation of the eyes, but by June he felt more hopeful of finishing the work. 'I only feel grateful to Him who dwells above the stars that I now begin once more to use my eyes', he wrote to his patron the Archduke Rudolph. 'I am at present writing a new Symphony for England, bespoken by the Philharmonic Society, and hope it will be quite finished fourteen days hence. I cannot strain my eyes as yet long at a time; I beg therefore Your Royal Highness's indulgence with regard to your Variations, which appear to me very charming, but still require closer revision on my part.' In the postscript of the same letter he mentions that in consequence of his illness and inability to write as much as usual, he had been obliged to incur a debt of from 200 to 300 florins, but hoped to recover this if his health improved. It is

THE PHILHARMONIC PERIOD

interesting to note that he borrowed money instead of selling his Austrian loan: Beethoven's attachment to gilt-edged securities had by this time become almost pathetic, but the accusation of miserliness that has been levelled at him needs qualification — Beethoven probably regarded his investments not as his own, but held in trust for his nephew against the time when Beethoven should die. At any rate, the very idea of parting with his shares invariably threw Beethoven into a panic.

The reaction of the Viennese nobility to the news that Beethoven's compositions were to have their first performance abroad was in keeping with their principles. They regretted his decision because it reflected discredit on Vienna. The spirit of patriotism to which Beethoven had appealed when applying for a post at the Court Theatre was as strong as ever, and a petition was drawn up and signed by the most influential citizens of Vienna, begging him to produce the *Mass in D* and the *Ninth Symphony* in their city, and to write a second opera which would prove that German opera could be superior to Italian. Naturally Beethoven was gratified to receive such evidences of appreciation; he had been endeavouring to obtain subscriptions for the *Mass* from a number of foreign rulers — in Prussia, France, Saxony, and Russia — but his intention had been to honour his agreement with the Philharmonic Society regarding the symphony. Under the influence of the Viennese appeal his resolution weakened; he allowed the symphony to be played at a concert in his honour and for his benefit at the Kärnthnerthor Theatre on May 7th, 1824, and three days later wrote to Probst, the Leipzig publisher:

> These are all I can at present give you for publication. I must alas! now speak of myself, and say that this, the greatest work I have ever written, is well worth 1,000 florins C.M. It is a new grand Symphony, with a finale and voice parts introduced, solo and choruses, the words being those of Schiller's immortal 'Ode to Joy', in the style of my pianoforte Choral Fantasia, only of much greater breadth. The price is 600 florins C.M. One condition is,

ENGLISH RELATIONS WITH BEETHOVEN

indeed, attached to this Symphony, that it is not to appear until next year, July 1825; but to compensate for this long delay, I will give you a pianoforte arrangement of the work *gratis*, and in more important engagements, you shall always find me ready to oblige you.

Beethoven then, having broken his word to the Philharmonic Society in response to an appeal to his patriotism, was seeking a publisher in Leipzig for his *Ninth Symphony*. Payment from the Philharmonic Society he had already received. The proceeds of the Vienna concert, however, were disappointing, for Beethoven received only 420 florins; a second performance was given on the 23rd with 500 florins guaranteed by the management of the theatre (who had eaten up the profits of the previous 'benefit') which resulted in a loss to the theatre occasioned by a sparse attendance. Offended first by the poor financial return and secondly by the lack of public support, Beethoven allowed his temper to get more than ever out of control; he invited to dinner those who had worked hardest to help him, and accused them over the table of having combined to cheat him. This got to the ears of the directors of the Philharmonic Society of London, and effectively scotched negotiations for a visit to London, under which Beethoven was to receive £300 for a symphony and a concerto, and a guaranteed benefit of £500. The rights of publication of the *Ninth Symphony* Beethoven sold to Schott of Mayence for 600 florins on July 19th, 1824, the dedication being to the King of Prussia. Not until 1825 did the Philharmonic Society receive the score of the symphony that had been promised exclusively to them for a date not later than March 1823. On the title page of the score appears the sentence: 'Geschrieben für die Philharmonische Gesellschaft in London.'

The *Ninth Symphony* had its first London performance on March 21st, 1825, conducted by Sir George Smart. A work of such dimensions, employing such a large orchestra, naturally presented considerable difficulty to those responsible for its performance. Sir George Smart had misgivings about his own

THE PHILHARMONIC PERIOD

understanding of the work, and wrote, a week before the concert, saying that he thought he could grasp the work but would prefer a postponement in the hope that Beethoven himself might be induced to come to London and conduct the work. Dragonetti, the famous double-bass player, said that had he seen the *Ninth Symphony* before naming his fee he would have asked double, and Dragonetti was a friend of Beethoven. Whatever reasons for grievance the members of the Philharmonic Society might have against Beethoven, they were swamped by the necessity of knowing how to obtain a true reading of the *Ninth Symphony*. Only Beethoven could enlighten them.

It is likely, therefore, that the 1825 performance of the *Ninth Symphony* was a bad one, but that matters little, for it was beyond the ken of the audience, anyway. Criticisms show how little it was understood even by the experts. *The Harmonicon* is typical of the general opinion of the time:

> In the present Symphony we discover no diminution of Beethoven's creative talent; it exhibits many perfectly new traits, and in its technical formation shows amazing ingenuity and unabated vigour of mind. But, with all its merits that it unquestionably possesses, it is at least twice as long as it should be; it repeats itself, and the subjects, in consequence, become weak by reiteration. The last movement, a chorus, is heterogeneous; and though there is much vocal beauty in parts of it, yet it does not, and no habit will ever make it, mix up with the first three movements. This Chorus is a Hymn to Joy, commencing with a recitative and relieved by many *soli* passages. What relation it bears to the Symphony we could not make out; and here, as well as in other parts, the want of intelligible design is too apparent.

Complete lack of appreciation of the main glory of the *Ninth Symphony* is the most outstanding feature of the above criticism. Beethoven's artistic feeling was towards a finale that would pick up the threads of the thematic material from the

ENGLISH RELATIONS WITH BEETHOVEN

three earlier movements, weave them into a new and original pattern that should reveal their relation to an idea greater than themselves, and on this finale build a climax into which a philosophic idea would emerge through musical means. For this purpose he needed voices and poetry. Beethoven so far outstripped classical conceptions of musical form that it is not surprising that both critics and performers were nonplussed. This was not the first impact of the new romanticism on London orchestral concert audiences, but it was the most uncompromising challenge that they had up to that time received.

Sir George Smart took the first opportunity to visit Beethoven in Vienna in order to obtain first-hand information about the interpretation of the *Choral Symphony*, but it was twelve years later before the Society, conducted this time by Beethoven's friend, Moscheles, essayed a second performance. This was better understood, but it is indicative of those times that the tenor soloist complained that it was 'most unreasonable for the Quartett to be asked to rehearse twice'. The 1855 performance conducted by Wagner set the seal on the true interpretation of the *Ninth Symphony*, although for other reasons some members of London's musical life were at that time unwilling to admit it.

Beethoven never came to London. Attempts were made to persuade him to come, but the dice were loaded heavily against him. His illness demanded long visits to Continental spas, and he was always suspicious of the unknown. Moreover, he could not have conducted his performances in London by reason of his deafness, and it is doubtful if his financial reward would have been comparable with that of Haydn or Spohr, whose personal appearances advanced their cause. Ries was no doubt giving good advice when he wrote to Beethoven from Godalming in 1825:

> I am glad that you have not accepted any engagements in England. If you choose to reside there, you must previously take measures to ensure your finding your

account in it. From the Theatre alone Rossini got £2,500. If the English wish to do anything at all remarkable for you, they must combine, so that it may be well worth your while to go there. You are sure to receive enough of applause, and marks of homage, but you have had plenty of these during your whole life. May all happiness attend you. Dear Beethoven, yours ever,

FERDINAND RIES.

Too many obstructions stood in Beethoven's way for life ever to be easy for him. His indomitable spirit soared above them in his music, but in daily life he was one of society's misfits. One could not love the later Beethoven for his personal traits, and his constant reiteration of his own virtues and lofty ambitions must have been particularly embarrassing to the English. In Vienna he had a large body of admirers, among them, no doubt, a good proportion of the type that would admire with equal felicity some strange and loathsome animal in the zoo, but in addition to these there were many who admired him for his music, symbolic of the new heroic spirit of romanticism: he symbolized the triumph of art over materialism, the imperishable nature of the human mind, and of German *kultur* in particular. As such Beethoven was admired and his material well-being guaranteed, despite the many objections to favourable treatment that his character raised.

To the English, Beethoven was a modern musician of the first rank and nothing more. What dealings they had with him were as a professional musician, and the English had tried to deal with him always as an honest professional man. They had not been called upon to treat him as an object of charity, although his infirmities were well known to them. Nowhere, however, were musicians more charitably disposed towards unfortunate members of their profession than in England. When, therefore, letters began to arrive from Beethoven stating that he was in urgent need of financial assistance, the English immediately turned all their sympathy in his direction.

Beethoven used a personal approach, as usual; Ries had now

ENGLISH RELATIONS WITH BEETHOVEN

left London, but Moscheles was there, and Sir George Smart had made Beethoven's acquaintance in Vienna when he went to obtain information about the *Ninth Symphony*. To both these men Beethoven wrote on February 22nd, 1827. To Sir George Smart:

> I remember that some years ago the Philharmonic Society proposed to give a concert for my benefit. This prompts me to request you, dear Sir, to say to the Philharmonic Society that if they be now disposed to renew their offer it would be most welcome to me. Unhappily since the beginning of December I have been confined to bed with dropsy — a most wearing malady the result of which cannot yet be ascertained. As you are already well aware, I live entirely by the produce of my brains, and for a long time to come all idea of writing is out of the question. My salary is in itself so small, that I can scarcely contrive to defray my half-year's rent out of it. I therefore entreat you kindly to use all your influence for the furtherance of this project; your generous sentiments towards me convincing me that you will not be offended by my application. I intend also to write to Herr Moscheles on this subject, being persuaded that he will gladly unite with you in promoting my object. I am so weak I can no longer write so I only dictate this. I hope, dear Sir, that you will soon cheer me by an answer, to say whether I may look forward to the fulfilment of my request.
>
> In the meantime, pray receive the assurance of the high esteem with which I always remain, &c., &c.

The letter to Moscheles runs:

> Dear Moscheles,
> I feel sure you will not take amiss my troubling you as well as Sir G. Smart (to whom I enclose a letter) with a request. The matter is briefly this: Some years since, the London Philharmonic Society made me the handsome offer to give a concert in my behalf. At that time I was not, God be praised! so situated as to render it necessary for me to take advantage of this generous proposal. Things are, however,

THE PHILHARMONIC PERIOD

very different with me now, as for fully three months past I have been entirely prostrated by that tedious malady, dropsy. Schindler encloses a letter with further details. You have long known my circumstances, and are aware how, and by what, I live: a length of time must elapse before I can attempt to write again, so that, unhappily I might be reduced to actual want. You have not only an extensive acquaintance in London, but also the greatest influence with the Philharmonic; may I beg you, therefore, to exercise it, so far as you can, in prevailing on the Society to resume their former intention, and to carry it soon into effect.

The letter I enclose to Sir Smart is to the same effect, as well as one I already sent to Herr Stumpff.[1] I beg you will yourself give the enclosed letter to Sir Smart, and unite with him and all my friends in London in furthering my object.

Your sincere friend,
BEETHOVEN.

In addition to these letters, Beethoven requested Schindler also to write to Moscheles and Cramer. The Society took immediate action to relieve Beethoven's distress. A General Meeting was summoned for February 28th, at which William Dance took the chair. Charles Neate moved, Jean Latour seconded, and the meeting unanimously carried: 'That this Society do lend the sum of One Hundred Pounds to its own Members, to be sent, through the hands of Mr. Moscheles, to some confidential friend of Beethoven, to be applied to his comforts and necessities during his illness.' Before the money could reach Beethoven he had again written to Smart and Moscheles. That to Sir George Smart is dated March 6th, and that to Moscheles March 14th; both are couched in much the same terms and contain further information about his illness and alarm for the future.

> I was operated on for the fourth time on the 27th of February, and now symptoms evidently exist which show

[1] The London harp-maker.

that I must expect a fifth operation. What is to be done? What is to become of me if this lasts much longer. Mine has indeed been a hard doom; but I resign myself to the decrees of fate, and only constantly pray to God that His holy will may ordain that while thus condemned to suffer death in life, I may be shielded from want. The Almighty will give me strength to endure my lot, however severe and terrible, with resignation to His will.
So once more, dear Moscheles, I commend my cause to you, and shall anxiously await your answer, with highest esteem. Hummel is here and has several times come to see me.
Your friend,
BEETHOVEN.

Remembering the Society's offer of 1817 and the affair of the *Ninth Symphony*, one may be tempted to say that the devil was sick; but Beethoven was not a devil. His was a life devoted to ideals very difficult to attain: in the process he evaded some of the responsibilities towards his fellow men that lesser men, not obsessed by unattainable visions, would have performed more conscientiously. Four days after writing the last letter he wrote again, for now the Society's gift had reached him. The letter is again effusive — almost embarrassingly so — but excusable under the circumstances:

No words can express my feelings on reading your letter of the 1st of March. The noble liberality of the Philharmonic Society, which almost anticipated my request, has touched me to my inmost soul. I beg you, therefore, dear Moscheles, to be my organ in conveying to the Society my heartfelt thanks for their generous sympathy and aid.
Say to these worthy men, that if God restores me to health I shall endeavour to prove the reality of my gratitude by my actions. I therefore leave it to the Society to choose what I am to write for them — a Symphony lies fully sketched on my desk, and likewise a new Overture and some other things. With regard to the concert the Philharmonic Society had resolved to give on my behalf, I would entreat them not to abandon their intention. In

THE PHILHARMONIC PERIOD

short, I will strive to fulfil every wish of the Society, and never shall I have begun any work with so much zeal as on this occasion. May Heaven only soon grant me the restoration of my health, and then I will show the noble-hearted English how highly I value their sympathy with my sad fate.

I was compelled at once to draw for the whole sum of 1,000 guilden, being on the eve of borrowing money.

Your generous conduct can never be forgotten by me, and I hope shortly to convey my thanks to Sir Smart in particular and to Herr Stumpff. I beg you will deliver the metronomed 9th Symphony to the Society. I enclose the proper marking.

Your friend, with high esteem,
BEETHOVEN.

George Hogarth, in his history of the first fifty years of the Philharmonic Society, summarizes the case for and against Beethoven's appeal to the Society. Hogarth fails, however, to realize Beethoven's wish to make provision for his nephew upon his death, and he also fails to take into consideration Beethoven's reluctance to risk a repetition of the stormy events that succeeded the benefit concerts in Vienna in 1824; and this is important, for it probably explains why Beethoven preferred to appeal to London rather than to Vienna. Apart from these failings, however, Hogarth's statement is to the point:

> But, though the illustrious musician died in circumstances of neglect and penury, which will ever reflect disgrace upon his country, and especially upon the great and wealthy capital in which he had spent almost the whole of his life, yet he was not in the state of absolute want which he had morbidly imagined. When the inventory of his effects came to be taken after his death, there were found, among some papers in an old decayed chest, Austrian bank bills to the value of about a thousand pounds in English money, besides the hundred pounds sent by the Philharmonic Society, which remained untouched. This discovery made no small noise in Vienna; and the public were, or affected to be,

much hurt by Beethoven's having applied for assistance of which he did not stand in need, and, what was worse, having applied to strangers in London instead of his friends and admirers in Vienna, by whom every necessary aid would have been promptly bestowed. But such clamours were idle and ridiculous. Beethoven, if not absolutely pennyless, was miserably poor. It was well known to his 'illustrious patrons' and his 'numerous friends and admirers' that he had for years been living in penury and denying himself of the 'common comforts of life'. And what, after all, did the accumulated savings of this life of poverty and privation amount to? The magnificent sum of eleven or twelve hundred pounds sterling, yielding the ample revenue of thirty or forty pounds a year! No wonder that Beethoven, only turned of fifty, with the probability of many years of life, and yet disabled from labour, looked with dread upon the prospect of destitution: he might have done so even if his mind had not been enfeebled by disease. As to applying to foreigners in London in preference to his friends and countrymen in Vienna, his doing so only showed the estimate he had been taught, by sad and life-long experience, to form of the value of their friendship.

WHY WEBER CAME TO LONDON

Beethoven's passion for freedom was in no way hampered by his lack of political understanding, for his freedom was of the mind. So, for that matter, was all German freedom except that of the ruling classes in Austria and Prussia. True, men still took pride in belonging to the Free Cities of the Rhineland, but their freedom had lost its economic basis when the overland trade route declined in favour of British sea transport. The products of German craft industries, too, came in for a bad time when cheap British manufactured goods began to pour out from the rapidly-expanding factories. Between 1815 and 1820 there was a boom in these goods as the money given by Britain to Austria, Prussia, and even some of the smaller German states, to woo them to the allied cause, came into circulation. It was an artificial boom from which Germany received no material benefit, but it paid fat dividends in art. The ruling princes, into whose pockets the money rolled, were able to re-establish their reputations as upholders of Western culture against the infidel Turk and the barbaric Slavs — for so the Germans traditionally regarded their eastern neighbours — and since French and Italian art were now almost as unpopular in Germany as Slav art, German romanticism advanced apace. Beethoven, in Vienna, linked his music to the thought of Goethe and Schiller, while Weber, moving as he did from state to state, was right in the midst of German romantic nationalism in its most active form.

The year 1815 saw Weber in Berlin at a time when Prussia was celebrating the downfall of French military power. Patriotic feeling was at its height, and Weber entered fully into the spirit of the times. Romantic secret societies, called *Burschenschaften*, sprang up, at which Nordic deeds of valour were sung and recited, and in university towns a poet-gymnast named Jahn became the figurehead of a romantic youth movement. Weber's settings of a cycle of songs entitled *Lyre and Sword*

WHY WEBER CAME TO LONDON

by Theodore Korner, a poet member of the *Burschenschaften*, spread through these societies all over Germany, and established his reputation wherever German freedom was lauded.

But the allegedly free Germany of the liberal intellectuals could not support Weber; like all intellectuals in Germany, he had to depend for a living on the favour of the aristocracy. He held office under many petty princes, and his experiences at their courts reveal the tragedy of German nationalism as well as its potential greatness. *Der Freischütz* was written while he held a post as Kapellmeister to the Dresden opera, but it was first performed in Berlin. Berlin, flushed still with victory, accepted the opera more readily than Dresden, for the King of Saxony had been slow in changing sides on the eve of Waterloo and had in consequence lost about half his kingdom. Sulkily he rejected as much of the new romantic opera as he dared, and Weber held his post at the opera house jointly with an Italian named Morlacchi. Weber therefore held office at a reactionary court where Italian Opera persisted more firmly than in all other German opera houses. His employment there was a face-saving device of a monarch who would otherwise have been unpopular with his middle-class subjects. Weber's greatest work for German Opera was done in Dresden, and received the support of the opera-going public, but it must not be supposed that he had any support from the lower classes, who were half-starved and unenfranchised. They counted for nothing on the side of the new free Germany.

Weber's health broke down and he was told that he could expect but a short continuance of life. If he took a year's complete rest in Italy he might hope to live for a further five or six years; otherwise the end would come quickly. Weber therefore found himself in a similar plight to Beethoven when he applied through Moscheles to the Philharmonic Society for assistance; Weber had no means of guaranteeing the welfare of his dependants after his decease, and, like Beethoven, Weber hesitated to appeal to the aristocracy. In early life he had done their bidding even to the extent of fraud and blackmail, and had

THE PHILHARMONIC PERIOD

allowed himself to be made the scapegoat for his employer's transgressions, but he had learnt his lesson, and had not forgotten the character of the German class of nobles. The German aristocracy were willing enough to ride to glory on nationalist steeds, but the fact that the steeds were bred in non-aristocratic stables could at any time be proved against them if their prancing went out of fashion. Weber, like Beethoven, turned to London in his distress. He entered into a contract with Charles Kemble, who was then in charge of Covent Garden Theatre, to provide music for a play on the theme of *Oberon*, to conduct twelve performances, and also conduct performances of *Der Freischütz* and *Preciosa*. For these services he was to receive £1000, and was to be free to accept other engagements in London. His doctor told him he was signing his death-warrant, which Weber knew to be true; but for the sake of his wife and family, and because he could think of no other means by which he could quickly earn so large a sum, he undertook the task. He came to London on March 5th, 1826; he died on June 4th. The labour of composition, the exertions of the journey, and the London climate all helped to bring about his speedy end, but a contributing factor was his conscience, which drove him to a last supreme series of efforts in which there can be seen no falling-off of his highest standard at a time when any kind of effort was a drain on his slender physical resources.

Weber's reception in London gratified him immensely. He was received with the greatest kindness by Sir George Smart, with whom he stayed, and with the greatest enthusiasm by Kemble and the Covent Garden audiences:

> Thanks to God and his all-powerful will I obtained this evening the greatest success of my life. The emotion produced by such a triumph is more than I can describe. To God alone belongs the glory. When I entered the orchestra the house, crammed to the roof, burst into a frenzy of applause. Hats and handkerchiefs were waved in the air. The overture had to be executed twice, as had also several

WHY WEBER CAME TO LONDON

pieces in the opera itself. At the end of the representation I was called on to the stage by the enthusiastic acclamations of the public; an honour which no composer had ever before obtained in England. All went excellently, and every one around me was happy.

Kemble had entered into negotiations with Weber on the strength of public interest in *Der Freischütz*, which had been a sensation in London in 1824; he was reasonably sure of a good return for his £1000. From the nobility alone does Weber appear to have lacked appreciation. There was some criticism from members of the Philharmonic Society on the artistic merits of the horn solo opening to the *Oberon* overture, but this was hypercritical, for there does not seem to have been any objection to the horn quartet in the overture to *Der Freischütz*, which is much more advanced in style. It may be said, in fact, that in this overture, and in *Preciosa*, where he uses eight horns in the orchestra and on the stage, Weber reaches the tip of achievement with this instrument prior to the all-important introduction of valves. Beethoven used four horns in his *Ninth Symphony*, crooked in two different keys, by which means an almost complete scale can be obtained and the use of notes that are out of tune in the natural harmonic series avoided; nor was Beethoven the first to do this — Mozart used four horns in some of his *divertimenti*, and the practice of crooking horns and trumpets in different keys in order to obtain a certain minimum of full chords in the brass was accepted by Haydn, Mozart, Beethoven, and Schubert as a matter of course, but Weber's apparent freedom of the horns is supreme.

But besides the effective use of French horns in the orchestra, Weber wrote a concerto for the instrument; he emancipated the viola from its modest position in the orchestra by writing for it two solo works, a set of six variations and an *Andante and Rondo Ungarese*. He also wrote a concerto for the bassoon, and two concertos and a concertino for the clarinet. Both the latter instruments had obtained much popularity in England among musicians in the humbler walks of life; they were much used in

THE PHILHARMONIC PERIOD

English churches until the coming of organs in the 'forties supplanted them. The clarinet's popularity increased most in the ten years succeeding 1810, which is the date of an advancement in the key mechanism. It is also the date of Erard's invention of the double-action harp, which had an immediate effect on society at a higher level. It became a favourite with young ladies (possibly because it showed off their arms to advantage) whose brothers kept their old respect for the flute. (It will be recalled that this was an eighteenth-century craze with well-to-do amateurs.) The combination of these two instruments guaranteed sweet harmony in the home for the greater part of the nineteenth century.

This was the London that Weber knew for so short a time. It lacked some of the features that Weber might have thought necessary in a fully-alive musical community. In particular the association of music with national cultural aspirations would not be obvious as it was in Germany. Unless he had some understanding of British political institutions he could not be expected to know why. In Germany progressive elements came up against autocratic monarchs, and therefore became nationalist, and strove to effect reforms by influencing the aristocracy. In Britain the monarchy ruled by favour of a Parliamentary system, and reformers therefore tried to influence Parliament, and, according to the fashion of their times, sought freedom by means of *laissez-faire*. Workers had to oppose *laissez-faire* by combining in trade unions, and these early trade unions organized themselves, like secret societies, with childish rituals for which many of their members were severely punished by law. It may be that the coolness of the British nobility towards Weber was caused by some suspicion that the *Burschenschaften* were secret societies of an anti-constitutional nature. Weber's benefit concert suffered from a possible misunderstanding, but it was left for his successor, Wagner, in 1855, to feel the full weight of London's political disapproval. Meanwhile, the cause of music went smoothly onward, gaining speed from the effects of lubrication by a German intellectual of a different type — Mendelssohn.

THE MENDELSSOHN TRADITION

The first half of the nineteenth century was above all the age of the pianoforte. Liszt made his first appearance with the London Philharmonic Society in 1827,[1] playing a concerto of Hummel, who was by this time recognized as the successor of Clementi in the development of pianoforte technique. With the arrival of John Field from St. Petersburg, in 1832, the charm of the Clementi style had a temporary revival, but the position of Hummel was never seriously challenged, and for years it was regarded as necessary for a pianist to make his first appearance with a Hummel concerto.

The day of the pianoforte recital had not yet arrived. In the late 'forties Charles Hallé instituted semi-private pianoforte concerts in his London house, but prior to this the programmes given by pianists were curiously mixed, and any pianist who wanted to give a concert had to engage vocal soloists and an orchestra. To their credit it must be stated that orchestral players were generally prepared to assist pianists at such concerts by playing without fee, but they would not in addition attend a rehearsal, so these events rarely had any artistic value. They were patronized by the parents and friends of the soloist's pupils, and were regarded as a means of advertising one's teaching practice rather than as serious contributions to art. Most pianoforte pupils were middle-class young ladies, with a sprinkling of clergymen, for the social scheme was very strict, and no other profession offering scope for artistic talent to young men of good breeding was free from social taint. It is a mistake to assume, however, that under these conditions pianoforte playing was in a bad way. The early nineteenth century papa was engaged often in a process of commercial and social entrenchment; putting his profits back into his business in order to build up his capital resources without un-

[1] His first London appearance was, however, in 1824.

necessary outside help, and planning for his family's social advancement. His daughters were often well drilled in the Clementi-Hummel school of pianism. Certainly, amateur pianists of to-day should beware of drawing comparisons.

The pianoforte concerto, therefore, remained the principal concert attraction for soloists. J. B. Cramer, Clementi, Moscheles, Cipriani Potter, John Field, Sterndale Bennett — so the line ran — and in this social background moved Mendelssohn, the peer of all.

Mendelssohn's introduction to the Philharmonic Society came through Attwood, and his first appearance in London, when he conducted his Symphony in C minor, was the outstanding success of 1829. That was the London point of view. Mendelssohn's account of the event is not so uncritical, but he had tact. In a letter to his sister Fanny he wrote:

> When I entered the Argyll Rooms for the rehearsal of my Symphony (May 24), and found the whole orchestra assembled and about two hundred listeners, chiefly ladies, strangers to me, and when, first, Mozart's Symphony in E flat was rehearsed, after which my own was to follow, I felt not exactly afraid, but nervous and excited. During the Mozart numbers, I took a little walk in Regent Street and looked at the people; when I returned, everything was ready and waiting for me. I mounted the orchestra and pulled out my white stick, which I have had made on purpose (the maker took me for an alderman, and would insist on decorating it with a crown). The first violin, François Cramer, showed me how the orchestra was placed — the furthest rows had to get up so that I could see them — and introduced me to them all, and we bowed to each other; some, perhaps, laughed a little, that this small fellow with a stick should take the place of their regular powdered and bewigged conductor — then it began. For the first time it went very well and powerfully, and pleased the people much, even at rehearsal. After each movement the whole audience and the entire orchestra applauded (the musicians showing their approval by striking their instru-

ments with their bows and by stamping their feet); after the Finale they made a great noise, and as I had to make them repeat it because it was badly played, they set up the same noise once more; the Directors came to me in the orchestra and I had to go down and make a great many bows. Cramer was overjoyed, and loaded me with praise and compliments. I walked about the orchestra, and had to shake at least two hundred hands. . . .
But the success at the concert last night (May 25) was beyond what I could ever have dreamed. It began with my Symphony: old François Cramer led me to the pianoforte like a young lady, and I was received with immense applause. The Adagio was encored — I preferred to bow my thanks and go on, for fear of tiring the audience; but the *Scherzo*[1] was so vigorously encored that I felt obliged to repeat it, and after the Finale they continued applauding, while I was thanking the orchestra and shaking hands until I had left the room.[2]

Better than anything this letter shows the spirit that pervaded the meetings of the Philharmonic Society. Its atmosphere was that of a professional club rather than a business enterprise. Although the Philharmonic orchestra as Mendelssohn found it was by no means perfect, it was better than any other London orchestra of that time, if we are to judge by Chorley and Hogarth's descriptions of other London orchestras. Hogarth said of the Society of Antient Concerts in 1835 that 'the orchestra, vocal and instrumental, embraces the greatest talent that can be obtained, and some of the most magnificent compositions of the last two centuries are heard at these concerts in all their grandeur. But their management is now liable to a charge of want of energy, activity and research. No trouble is taken to bring to light the innumerable gems that lie in the vast stores of ancient music: and the performances consist of little more than a repetition of a few pieces of Handel and a few

---

[1] The *Scherzo* was an interpolation. Arranged from that of Mendelssohn's Octet.
[2] An interesting letter. Mendelssohn had a baton, but Cramer led him to the pianoforte. Did the old system of dual control survive as a ritual of introduction?

THE PHILHARMONIC PERIOD

other masters, which have become familiar to everybody who has the slightest knowledge of music. This is a perversion of the object of the institution, and is mere waste of the ample means at its disposal'. Chorley's comments on the Italian Opera[1] at the time are equally caustic: 'The orchestra was meagre and ill-disciplined; the chorus was an ear-torment rather than an ear-pleasure; the scenery and appointments were shabby to penury.'

Credit for the improvement of the opera orchestra goes to Costa, who took charge in 1832, but Costa could have done nothing had not the economic situation changed in favour of the musical profession. In the 'thirties opera had lost much of the social exclusiveness that it had enjoyed in the Regency, but it was still dependent on people living permanently in London. Operatic performances took place on Tuesdays and Saturdays only, with a 'long night' on Thursdays catering for a 'popular' public. The railways changed all that, making it possible for provincial patrons to attend the opera; in consequence opera came to be performed more regularly, but over a shorter annual season, and casual attenders grew more numerous than regular subscribers.

This change allowed Costa to offer more regular employment to his orchestra, and to improve their standard of performance by constant rehearsal, a change from which the Philharmonic Society must have derived benefit. The directors appear not to have taken this long view, however, for they complained publicly on several occasions of the opera prohibiting singers and players from attending Philharmonic concerts because they were required to be at the opera. It seems strange that a society having so much the atmosphere of a professional club should not have given more thought to performers' economic problems, but even to this day it is so. There was not, then, in the early nineteenth century any permanent employment for players at symphony concerts — all such orchestras were recruited from players engaged mainly in theatre bands or in private teaching.

[1] H. F. Chorley, *Thirty Years' Musical Recollections*, 1862.

THE MENDELSSOHN TRADITION

Discipline therefore was lax, and the players are said to have been ill-mannered; even fifty years after the society's foundation these players were little better as musicians, though their social status had improved. This social improvement was brought about mainly by economic developments (though these left much still to be desired), but the personal influence of Mendelssohn, Liszt, and not least, of the Prince Consort, influenced public opinion towards a more favourable acceptance of musicians in society.

How, then, did Mendelssohn, a young German, play so great a part in the improvement of the prestige of his art in a foreign country? We have already noted that the English looked abroad for guidance in musical taste, and that there was in the early nineteenth century a strong feeling away from Italian styles towards German styles; it was natural therefore that any German musician of outstanding merit visiting our shores should be received with more favour, perhaps, than was his due. But in order to wield a permanent influence it was necessary for the German to reciprocate some of the feeling shown to him. This was possible by one who felt that his life was likely to be bound up with the English. Haydn was grateful to the English, but he could not have brought himself to live among us. Weber came for money, and died before he really got to know us. The only German who knew us sufficiently well to have become a leader of opinion here (besides Mendelssohn) was Spohr, whose compositions were treated with respect but not felt to be of permanent value. Criticism was to the point. Spohr's *Die Weihe der Tone* had to justify the apparent absurdity of representing by means of sound the deep silence of nature before sound existed, and critics seized on this point. Similarly, they attacked his *Historical Symphony* which set out to describe the historical trend of music from 1720 onwards. 'It is useless to give imitations of Bach, Handel, Haydn, Mozart and Beethoven, when their actual works are well known: and the introduction of such various styles into one work must render it patchy and incoherent.' Nor did Spohr fare any better when he delved into metaphysics for themes for his programme symphonies. The symphony *Descriptive of the Conflict of Virtue and Vice in Man* ought to have made a sentimental appeal to Victorian audiences, but the critics would have none of it. 'It is evident, from the mere mention of the subject of this symphony, that Spohr, with all his knowledge of the philosophy of his art, mistook, in this instance, its powers and objects, endeavouring to employ it in the expression of abstract ideas and moral sentiments, to which musical sounds have no greater analogy than to the demonstration of a proposition of Euclid.' This symphony revived the style of the *concerto grosso* — a *concertino* of eleven players alternating with a large *ripieno* — but the effect was said to be in no way different from effects common in ordinary orchestral playing. Spohr's theories were politely received but never came to be loved: Mendelssohn's melodies were loved, and whatever extraneous associations were implied by his titles gained approval because they added interest to, but did not strive to justify, the music itself.

Mendelssohn's behaviour won the hearts of the English people, too. He came to this country first at the age of twenty in consequence of an agreement with his father under which he was to travel widely through Europe, observing the musical life of the people, with the object of deciding where he should settle and pursue his chosen profession as a musician. So his attitude towards the English was sympathetic and tactful. In all things he sought to appreciate the English point of view, and this in itself tended to recommend him to the English in a way that Beethoven would not have understood. In the ordinary course of his relations with the Philharmonic Society Mendelssohn had to deal with proposals similar to those Beethoven had encountered, but where such proposals had been openings for discussion with Beethoven, with Mendelssohn they were introductions to a career, and the promises he made were kept. Indeed, Mendelssohn did more: when in 1832 the Philharmonic Society offered him a fee of a hundred guineas for a symphony, an overture and a vocal piece, on the under-

THE MENDELSSOHN TRADITION

standing that the Society should have the right to perform these works at all times, but that copyright should revert to the composer after the expiration of two years, Mendelssohn's reply so impressed the directors of the Society (who no doubt remembered Beethoven's attempt to bargain over a much better offer in 1817), that it deserves to be quoted:

> I beg you will be so kind as to express my sincerest acknowledgements and my warmest thanks for the gratifying manner in which the Society has been pleased to remember me. I feel highly honoured by the offer the Society has made, and I shall compose, according to the request, a Symphony, an Overture, and a vocal piece under the conditions mentioned in the resolution. When they are finished, I hope to be able to bring them over myself, and to express in person my thanks to the Society. I beg that you will let me know whether my compositions are expected to be ready for the next session, or whether the arrangements for it are already complete without them. At all events, I shall lose no time, and I need not say how happy I shall be in thinking that I write for the Philharmonic Society.

Mendelssohn's plan for a musical career had begun by this time to fall into shape. Perhaps it is indicative of his Jewish ancestry that his plan should be so well laid, so carefully tested, and so conscientiously performed. In its broadest outline it was suggested by his father, but with infinite wisdom the details were to be decided by Mendelssohn himself after a period of practical experience. How well the plan was working out, even as early as 1832, is revealed in a letter from Felix to his father on February 21st of that year:

> It is now high time, dear Father, to write you a few words with regard to my travelling plans, and on this occasion in a more serious strain than usual, for many reasons. I must first, in taking a general view of the past, refer to what you designed as the chief objects of my journey; and wished me strictly to adhere to. I was closely to examine the various countries and to fix on the one where I wished to live and

THE PHILHARMONIC PERIOD

work; I was further to make known my name and capabilities, in order that the people among whom I resolved to settle, should receive me well, and not be wholly ignorant of my career; and finally, I was to take advantage of my own good fortune, and your kindness, in preparing the ground for my future efforts ... It is a happy feeling to be able to say, that I believe these objects have been carried out. Always excepting those mistakes which are not discovered till too late, I think I have fulfilled these purposes. People now know that I exist, and that I have a purpose, and any talent that I display, they are ready to approve and accept. They have made advances to me here[1] and asked for my music, which they seldom do; as all the others, even Onslow, have been obliged to offer their compositions. The London Philharmonic have requested me to perform something new of my own there on the 10th of March. I also got the commission from Munich without taking any step whatever to obtain it, and indeed not until *after* my concert. It is my intention to give a concert here (if possible), and certainly in London in April, if the cholera does not prevent my going there: and this on my own account, in order to make money, that in this respect also, I may have felt my way before returning; I hope, therefore, I may say that I have also fulfilled this part of your wish — that I should make myself known to the public.

Your injunction too, to make choice of the country that I preferred to live in, I have equally performed at least in a general point of view. That country is Germany.

It is all so natural, Mendelssohn was not only finding full expression of his artistic merits but was at the same time finding happiness. To those who showed interest in his music he reciprocated by considering their interests. The performance on May 10th, referred to in the above letters, did not take place, but *Fingal's Cave* had its first performance on May 14th of that year, and Mendelssohn personally played the solo part in his *Pianoforte Concerto in G minor* on May 28th. The composer's appreciation of the Philharmonic Society was shown by his

[1] i.e. Paris, whence the letter was sent.

offer to present the manuscript score of *Fingal's Cave* to them, an honour the Society acknowledged by presenting the composer with a piece of plate. Time had been when Leopold Mozart had complained of the habit that induced the gentry to present him with jewelled swords and snuff-boxes instead of a fee, but by Mendelssohn's time the status of the first-class musician had so altered that an exchange of gifts between him and the concert promoters was possible on a basis of social equality without interfering with their right to conduct future arrangements in a strictly business manner. Indeed, by November of the same year the Society was making to Mendelssohn the offer of a hundred guineas for three compositions already quoted on page 152.

Again, Mendelssohn's reaction was to give his patrons more than they had asked for. 'I beg you will inform the Directors of the Philharmonic Society that the scores of my new Symphony and Overture are at their disposal', he wrote on April 27th, 1833, 'and that I shall be able to offer them a vocal composition in a short time hence, which will complete the three works they have done me the honour to desire me to write for the Society. But as I have finished two new Overtures since last year, I beg to leave the choice to the Directors as to which they would prefer for their concerts; and in case they should think both of them convenient for performance, I beg to offer them this fourth composition as a sign of my gratitude for the pleasure and honour they have again conferred upon me.'

The effect was still further to enhance Mendelssohn's standing with the Society. He had had a bad winter in Paris, some of his best friends there died of the cholera, and he himself had been severely ill with the same disease. It was a relief to be again in London, where fortunately this scourge of nineteenth-century city life was for a time not greatly in evidence, and to see again his London German friends, Klingemann, Rosen, and Moscheles. The English, too, made much of him.

> I must really describe one happy morning last week; of all the flattering demonstrations I have hitherto received, it is

THE PHILHARMONIC PERIOD

the one which has most pleased and affected me, and perhaps the only one which I shall always recall with fresh pleasure. There was a rehearsal last Saturday at the Philharmonic, where however nothing of mine was given, my overture being not yet written out. After Beethoven's *Pastoral Symphony*, during which I was in a box, I wished to go into the room to talk to some friends, scarcely however had I gone down below, when one of the orchestra cried out, 'There is Mendelssohn!' on which they all began shouting and clapping their hands to such a degree that for a time I really did not know what to do; and when this was all over, another called out 'Welcome to him!' on which the same uproar commenced, and I was obliged to cross the room, and to clamber into the orchestra, and return thanks.

Never can I forget it, for it was more precious to me than any distinction, as it showed that the *musicians* loved me, and rejoiced at my coming, and I cannot tell you what a glad feeling that is.

So Mendelssohn's cause increased in influence, but it was not always done without opposition: players there were who resented his popularity, and in the Philharmonic Society it was possible for them to express their opinions. However much we may admire the principle of freedom of opinion, it must be confessed that the temptation to penalize those of an opposite opinion is often hard to resist. At the first performance of *Elijah* at the Birmingham festival of 1846 it came to Mendelssohn's notice that certain London orchestral players had not been engaged because they had been heard to criticize Mendelssohn's music adversely. Mendelssohn at once let it be known that such action was repugnant to him, and that players ought not to be penalized for their opinions. Within ten years controversy on musical taste was to become almost violent, with Wagner openly showing his disgust at Mendelssohn's music, even from the Philharmonic platform, but time is a great teacher — we can admire Mendelssohn's fairness while we are obliged to find excuses for Wagner; the irony of it is that we are

THE MENDELSSOHN TRADITION

apt to excuse Wagner because he was a foreigner, claiming the qualities of fair and balanced judgment as British qualities. That they are by no means peculiar to the British is obvious from the behaviour of Mendelssohn, and Haydn too; but because he had so many of those qualities that the British, at least in theory, admired, Mendelssohn established a reputation in this country that had an enormous influence on our music.

This influence did little to advance orchestral technique. Mendelssohn was not a reformer but an advanced conservative in music. He brought an urbanity into music that came as a useful foil to the rugged grandeur of Beethoven, and he brought a polish to orchestral playing that was all to the good. With the polishing process certain developments took place that had to be corrected later, such as a general rise in the pitch of English orchestral instruments, and a tendency for tempos to quicken in standard classical works; these developments took place when the Mendelssohn tradition was all-important in London orchestral circles, but they were not Mendelssohn's own innovations — rather were they brought about by conductors eager to attract attention to their own interpretations of well-known works. Mendelssohn was symbolic of his age: he strove to bring to perfection in music the artistic thought of his contemporaries. Some of this was unduly ornate — as was only to be expected of a social set growing to affluence as a result of expanding trade — but much of it aimed at a standard of genteel beauty that belongs to no other age but the Victorian. In the process of improving their social conditions, as they acquired more wealth and power, the middle classes favoured a type of art that shunned the grimmer side of life from which they were seeking to escape, and sought a carefree state of mind wherein all was innocuously beautiful and redolent of the most common Christian platitudes. Mendelssohn accepted and indeed sublimated their point of view; he was able to run a close second to Handel in the popularity of his oratorios, he was superior to Clementi and Hummel in his pianoforte works, he could not outstrip Beethoven in the symphony, the concerto, nor in chamber music, but he commanded a respect for the personal honour of the professional composer that Beethoven had previously strained. In Mendelssohn the spirit of the Victorian ideal finds its most perfect expression in music; later generations have questioned its completeness, but an appreciation of Victorian greatness — and the Victorians were undoubtedly great — would be impossible without an appreciation of the greatness of Mendelssohn.

THE ORCHESTRA IN LOW LIFE

MENDELSSOHN made his appeal to two classes of musical taste. The higher of these formed the audience of the Philharmonic Society's concerts, while the second — a much larger class — was interested in oratorio. Religion held together the greatest mass of opinion in the nineteenth century, and little progress could have been made by any musical body that ignored the demand for works with a Christian flavour. The Philharmonic Society had been founded to foster orchestral music, and oratorio therefore the Society rarely attempted. When it was attempted, the facilities for its presentation open to the Society were inferior to those enjoyed by that great oratorio-performing organization the Sacred Harmonic Society, which employed a chorus and orchestra of 300 in 1836, rising in number to 700 by 1848 when Costa became conductor. They held their meetings in Exeter Hall, but it is indicative of the gulf between the two classes of musical taste that when in 1836 the directors of the Philharmonic Society approached the Exeter Hall authorities for permission to use their hall, the authorities — an influential nonconformist body — before they would make a decision, asked the directors if there were any immoral tendencies in the music they wished to play.

The Philharmonic Society was in fact like a little island of carefully-chosen music surrounded by a lagoon of popular oratorio beyond which the ever-swelling tide of common entertainment beats with increasing force on the hard reefs of a well-defined artistic barrier. It was not even easy for some of the oratorio worshippers to become members of the Philharmonic Society; it is on record that opposition was offered to one who sought membership of the Society, but his proposer succeeded by explaining that although it was true the nominee was a tradesman, nevertheless he did not serve behind the counter. But between the great oratorio-loving public and 'the mob' an

THE PHILHARMONIC PERIOD

even stronger barrier had to be maintained, for the middle classes were beginning to wonder what would happen if ever 'the mob' got out of hand. 'The mob' was a serious problem — quite as serious as the prevalence of cholera or the disturbing fluctuations in the price of corn — and slowly it was beginning to dawn on middle-class thought that these problems could not be entirely blamed on the devil or be attributed any longer to the wrath of God. The humanitarian politicians and nineteenth century humanitarian novelists have their places in British history, but the humanitarian music-lover has been overlooked. When George Hogarth wrote his *Musical History, Biography and Criticism* in 1835, he ended his book on a hopeful theme:

> The diffusion of a taste for music, and the increasing elevation of its character, may be regarded as a national blessing. The tendency of music is to soften and purify the mind. The cultivation of musical taste furnishes to the rich a refined and intellectual pursuit, which excludes the indulgence of frivolous and vicious amusements, and to the poor, a *laborum dulce lenimen*, a relaxation from toil, more attractive than the haunts of intemperance.

Together with the haunts of intemperance Hogarth had in mind must be classed the theatres and pleasure gardens, in which there was an enormous amount of music, mostly of a commonplace order. Prior to the passing of the Theatres Act in 1843, only three London theatres — Covent Garden, Drury Lane, and the Haymarket — were allowed to produce 'legitimate' drama (i.e. five-act drama without music). These were known as the major theatres. The minor theatres were obliged to render their plays 'illegitimate' by the introduction of songs and accompanying music. Even Shakespeare's tragedies were performed in this way, the shows advertised as 'burlettas'. These 'burlettas' often proved more popular than 'legitimate' plays and the major theatres accordingly employed music whenever it suited their policy. The only one of the three major theatres that conscientiously followed a 'legitimate' policy seems to have been the Haymarket Theatre. (This

THE ORCHESTRA IN LOW LIFE

theatre must be distinguished from the opera house in the Haymarket, called the King's Theatre.) Opera had always been dependent on wealthy patronage, but by 1830 it was in a very bad way. In that year the Philharmonic Society, in consequence of a fire at the Argyll Rooms (where their previous concerts had been held) moved to the King's Theatre, but their experience was unfortunate: within a short time of this arrangement letters of complaint began to be received by the secretary — some of them anonymous, for the subjects of complaint were often indelicate — with the result that the theatre management invited a committee from the Philharmonic Society to consult with the stage manager, and practically promised that whatever alterations the committee desired should be carried out, if only the Philharmonic Society would continue to use their premises. The Society remained at the King's Theatre until 1833, but the premises were still most unsatisfactory, and the Society therefore moved to the old Hanover Square Rooms that Gallini, Bach, and Abel had built in 1774. The rooms had a reputation for excellent acoustical properties, but the theatre's concert-room held a much larger audience, and it is possible that the Philharmonic Society might have increased its membership had more seating accommodation been available, without necessarily lowering the social status of the audience.

There could be no compromise, however, between a society like the Philharmonic and the theatres as they were then being run. The minor theatres were quite outside the pale of respectability. The plays given there were often crude and the audiences cruder. The buildings were evil-smelling and ill-lit, their corridors a happy hunting ground for prostitutes, while parasites of a humbler biological order infested the seats of the auditorium. Vendors of various delicacies — hot saveloys, bacon sandwiches, fried fish, pots of beer and glasses of gin and brandy — passed round the auditorium whilst the play was in progress. This, it may be said, was no background for serious music; but managers had noticed that their audiences responded to

THE PHILHARMONIC PERIOD

music. Hogarth's theory might be made to work. The problem had been tackled in France, where popular education was more pronounced and the factory system not exploited to the inhuman extent common in England. There Philippe Musard had succeeded in gaining popularity for promenade concerts in 1833. To some extent this popularity was based on the attraction of a new instrument called the cornet-à-pistons, but dance music was also a prominent feature of Musard's programmes. London theatre managers found it profitable to stage 'Promenade Concerts à la Musard', at which the music was entirely instrumental, and the seats of the auditorium were boarded over to hold a standing audience. The Lyceum led the way in 1838 with an orchestra of sixty conducted by Negri. His programmes were made up of four overtures, generally French, four quadrilles, mainly by Musard, a cornet or wind-instrument solo, and four waltzes by Strauss or Lanner. A rival show at the Crown and Anchor Tavern in the following year introduced symphonies in the first half of the programme, but Valentino, the conductor (a musician of advanced taste with a long record for useful work at the Paris Opera), failed to make his London venture pay. At this point one of the 'major' theatres, Drury Lane, took up the challenge, and in 1840 instituted *Concerts d'été* under the direction of an English violinist named Eliason with the assistance of a Frenchman named Jullien. The Lyceum responded by bringing over Musard in person, who appeared at that theatre during October, November, and December in that year. He has passed into English literature in an unworthy rhyme by Thomas Hood:

> From bottom to top
> There's no bit of the Fop,
> No trace of your Macaroni;
> But looking at him,
> So solemn and grim,
> You think of the Marshals who served under Boney.

The amount of competition in this type of entertainment during that year makes the picture historically confusing, but

THE ORCHESTRA IN LOW LIFE

indicates considerable public demand. Eliason employed a band of ninety-eight players and a small chorus of twenty-six singers. The picture becomes clearer in the following year, however, when Jullien took command at Drury Lane with a band of ninety and a chorus of eighty. He reverted to Musard's original title of *Concerts d'hiver*.

Before Jullien came to London in 1840 he had had some experience as a conductor of dance music at the Jardin Turc, in Paris, and as a composer — or rather compiler — of dance music. His reputation was for quadrilles on topical tunes or with topical titles. In 1838, however, he had become insolvent, and fled to London. Here he succeeded where Valentino had failed, for despite reduced prices of admission, the venture was financially successful, and Jullien continued to give orchestral concerts at various theatres in London each winter until 1859, always using a large orchestra. In between his London concert sessions he took the whole of his organization on tour round the provinces, a task involving enormous difficulty with the travelling facilities then available. Certainly Jullien, whatever his faults, was not afraid of work.

His performances were a seven days' wonder — Jullien did not want them to thrill the world any longer than seven days, for he wanted another audience the following week. He kept the programmes topical by producing each season a new monster quadrille, generally on some patriotic theme, which was the main attraction of his programme and carried the audience through the rest of the evening's fare, not all of which was equally trivial. That is the justification for Jullien; all fair-minded critics had to admit that he brought occasionally some really worth-while compositions to the notice of a public that otherwise would not have heard them. Typical of many is this letter from one of Jullien's players, written long after Jullien was dead, to the music critic Joseph Bennett:

> Sir — With reference to your notice of the Promenade Concert, Queen's Hall, in this day's issue, I beg to remind you that in days gone by a Beethoven programme at a

Promenade Concert was deemed a bait likely to attract a large audience, and never was Covent Garden or Drury Lane more crammed than on a Beethoven night, when Jullien gave the concerts. I have no desire to appear in print, but I do not think that the great work of a man who was, I believe, the first manager and conductor to familiarise the public with classical orchestral works, should be forgotten. Moreover I may say that I never heard finer performances of many classical Overtures and Symphonies than those directed by Jullien, and I have played them under all the great(?) conductors for over fifty years. Jullien, of course, was considered a charlatan by all who did not, or would not, understand him; but the twenty or thirty classical works he had made a study of, no one I have known has made go so well.—With apologies, I am, yours sincerely,

JOHN REYNOLDS.

It is true that under the baton of Jullien some of the works of Beethoven were first brought before a popular audience, but it would be misleading to imply that this was his most permanent contribution to music in England. He was a master of showmanship. Jullien conducting his Monster Quadrille was a godsend to contributors to *Punch*; his coat-tails flew about his legs, his hair fell about his eyes and his long moustachios twitched as he worked himself into a frenzy that culminated in his seizing a violin or a piccolo and adding to the din of the climax, after which he would flop prostrate on an elaborate velvet chair placed there for the purpose, his flashily-embroidered shirt front splendid in the glare of the footlights, and his extraordinary long cuffs dangling down over his hands; but when he conducted Beethoven Jullien used a special jewelled baton and spotlessly white kid gloves, brought to him on a silver salver. He restrained the raptures of the audience with gestures at once dignified and magnanimous, while he continued to direct his 'mass of executive ability' with the other hand. *Punch* called him 'The Mons', and 'The Mons' for eighteen years was the nineteenth-century equivalent of Shake-

'MANNERS AND CUSTOMS IN OLDE ENGLAND, IN 1849. A PROMENADE CONCERT.' From *Punch*

THE PHILHARMONIC PERIOD

speare's Jacques — he used his folly as a stalking-horse, and under cover of it he shot, not his wit, but his own interpretation of the classics.

Nevertheless, Jullien failed in the one task which he could have accomplished with more credit than any other London conductor — the popularization of Berlioz. Jullien brought the works of Beethoven before a new audience, but it is unlikely that his interpretations were more authentic than those of Beethoven's friends in the Philharmonic Society. Berlioz, however, he could have interpreted well. In 1838 Antonio James Oury wrote from Paris suggesting that the Philharmonic Society should do its utmost to secure the presence and a performance of the works of Berlioz, whom he described as 'the living Beethoven'; nothing happened immediately, but it appears that a score of the overture to *Benvenuto Cellini* was sent to Cipriani Potter, who had it in mind to conduct this work. Potter's courage failed him, however, when he saw the dimensions of the score, and no performance took place until 1841, when Charles Lucas conducted. Lucas was an undistinguished man who seems to have spent most of his life stepping into posts that the almost equally undistinguished Potter vacated, and the performance of Berlioz's overture was coldly received. No further performances of Berlioz's work were heard at the Philharmonic until 1853. The opportunity was open therefore for Jullien to exploit the possibilities of Berlioz in London; yet it was in his association with Berlioz that Jullien crashed.

It all came through that fatal fascination that opera has for the ambitious. In 1847 Jullien leased Drury Lane theatre, engaged Gye as manager, Berlioz as conductor, and a host of officials in every possible capacity; Sir Henry Bishop, for example, was engaged as 'inspector-superintendent at rehearsals'. Everything was on the most lavish scale, and unquestionably of the finest quality. Sims Reeves made his first appearance in opera on the opening night, singing in Donizetti's *Lucia di Lammermoor*. Berlioz's fee as conductor was to be 10,000 francs, with a further 10,000 francs for the expenses of four concerts

THE ORCHESTRA IN LOW LIFE

and a commission to write a three-act opera for the second season. Needless to say the much-disappointed Berlioz had accepted the post with pleasure: 'Art in France is dead', he wrote to his friend Ferrand, 'and one must go where it still lives.' *Lucia di Lammermoor*, in spite of the excellent production, was not a success, nor did the same composer's *Linda di Chamauni* fare any better. Balfe's *Maid of Honour* followed, and Mozart's *Figaro*, without any improvement in the situation; then Berlioz put on Gluck's *Iphigenia in Tauris* and Jullien fled to the Provinces to raise money, leaving an unpaid Berlioz to deal with an unpaid company. How Jullien could possibly have expected to make such a concern pay is a mystery; he had always previously made profits out of daring ventures, and seems to have been of the opinion that so long as large forces drew large audiences financial success must follow. Berlioz's hopes suffered, and we do not hear of him again in England until the great Exhibition of 1851, when he acted as a member of the jury and wrote an admirable report on the musical instruments submitted for competition at this event.

Most important was an invention of Dr. J. P. Oates for an improved type of piston valve for brass instruments. Dr. Oates' entry gained a prize medal at the Exhibition and revolutionized the brass section of the modern orchestra. At the same Exhibition Adolph Sax displayed his newly-invented saxhorns and saxophones, together with a bastard type of instrument called the saxo-tromba, that has not survived. The saxhorn family quickly gained popularity among the working men of Lancashire who, even as early as 1821, had shown a fondness for brass bands. These men visited the Great Exhibition at Hyde Park in 1851, in answer to a general appeal issued by the promoters, the response to which went far to change London opinion on the nature of 'the mob'.

> The state of the Metropolis throughout the whole period of the Great Exhibition will be remembered with wonder and admiration by all ... Instead of confusion, disorder

THE PHILHARMONIC PERIOD

and demoralization, if not actual revolution, which were foretold by some gloomy minds; instead of famine and pestilence confidently predicted by others, London exhibited a wonderful degree of order, good-humoured accommodation of her crowds, and power to provide for their wants ... Enormous excursion trains daily poured their thousands ... Throughout the season there was more of unrestrained and genuine friendship, and less of formality and ceremonial than has ever been known. It was like ... a gigantic picnic ... large numbers of work-people received holidays for the purpose ... 800 agricultural labourers in their peasants' attire from Surrey and Sussex, conducted by their clergy, at a cost of two and twopence each person — numerous firms in the North sent their people, who must have been gratified by the sight of their own handiwork — an agricultural implement-maker in Suffolk sent his people in two hired vessels, provided with sleeping berths, cooking apparatus, and every comfort ... which were drawn up to a wharf in Westminster, and furnished houses to the excursionists — a foreman was there to enforce the rules.[1]

The period was one of some educative value to Londoners, who were familiar with working-class political movements as mobs coming to Westminster with petitions. In the North 'the mob' formed a dangerously large part of the community, and oppressive measures were difficult to maintain: for some time tired and frightened employers had found it expedient to encourage working-class schemes for mutual improvement, especially when associated with religious life and music. This support, however, was not so readily given to the many workers who preferred brass bands, for brass bands had been known to lead processions of strikers, and their centres of attraction were more convivial than religious. Ale made men reckless, but religion taught resignation, so employers were in favour of religion and strongly opposed to 'haunts of intemperance'.[2] Appeals to employers for financial assistance to-

[1] Charles Tomlinson, *Cyclopaedia of Useful Arts*.
[2] Bands often took on the name 'Temperance Band' but only nominally did their members 'temp'.

THE ORCHESTRA IN LOW LIFE

wards the purchase of brass band instruments were rarely successful until late in the century, when the advertisement value of bands bearing the name of some colliery or factory began to be realized. The Great Exhibition of 1851, however, proved an inspiration to the northern brass band enthusiasts, and to one John Jennison, the owner of Belle Vue, Manchester. From 1853 onwards the annual competitions for brass bands held at Belle Vue have been events of great importance in the north, to which the Crystal Palace competitions, started in 1860, have been events no less fiercely contested. The 1853 meeting at Belle Vue can be clearly related to a demand from the members of working-class amateur bands who had assembled at the Hyde Park Exhibition of 1851, and the influence of Jullien can be seen in their choice of pieces from that time onwards; French overtures predominate.

Jullien had an even greater love of noise than Berlioz: he used as many as six brass bands to add to the effect of his Monster Quadrilles at their most extravagant period. But while the Monster Quadrilles were the high-spot of the programmes, and symphonies were drawn on to provide ballast for Jullien's otherwise light cargo, the main portions of his programmes were of standard French overtures — works which Jullien was able to interpret well. Auber, Boieldieu, Herold, and Adam were Jullien's standbys, and it is these works which the brass bands adopted and continued to perform well into the twentieth century when symphony orchestras had outgrown them. Here is Jullien's claim to have influenced posterity — not in his Monster Quadrilles, which are forgotten, nor in his performances of Beethoven symphonies, which count for little beside the steadily-pursued policy of the Philharmonic Society.

Jullien, however, took a lot of the stiffness out of the type of audience that had grown up with nineteenth-century orchestral music. He could not be other than comical, even after the worry of financial distress had begun to wear him down. One of his favourite stories, repeated *ad nauseam*, dealt with his own misfortune: 'I take Covent Garden', he would say, 'and give what you call Promenade Concerts. Mr. Gye, he come for rent on foot. Another season I make less money: Mr. Gye, he come for rent in a cab. Figure to yourselves, my friends: I give another season; Mr. Gye, he come for rent in a carriage. I make nothing! Name of a drummer!'

There is much truth in it, but in Jullien's case the cupidity of landlords was but a small cause of loss in comparison with his megalomania. The day of the great orchestra was coming, but Jullien strove to bend it to a trivial theme. He died in a madhouse.

Yet his intentions were good. He believed in giving the public only the best performers: Ernst, Sivori, Bottesini, Wieniawski, Sainton; Arabella Goddard, Marie Pleyel, Charles Hallé, Vinnier; Sims Reeves, Pischek, and many another fine artist appeared before audiences who could not possibly have afforded half-guinea tickets. Jullien's Promenade Concerts were a new and important development in concert promotion.

COSTA AND BERLIOZ

Not the least among the attractions offered to those visiting London for the Great Exhibition were the oratorio performances of the Sacred Harmonic Society. This energetic amateur society gave no less than thirty-one weekly performances between May 1851 and September of that year, singing alternately Handel's *Messiah*, Mendelssohn's *Elijah*, and Haydn's *Creation*, under the baton of Michael Costa. They were magnificent performances of their type, employing forces nearly seven hundred strong, remarkable not for any idiosyncrasies of interpretation, but for a solid round tone in the chorus and no humbug in the orchestra.

That was what the people wanted. At heart the British people are a race of craftsmen rather than of artists, and a good craftsman is not so much concerned with thinking out new ideas as with doing a known thing better and better until he can do it supremely well. That spirit dominated choral music during the nineteenth century — it is still by no means dead — and explains why the development of choral technique depended so much on competitive festivals. Costa's oratorio performances were old-fashioned by the end of his life, for until the last he employed an ophicleide in the orchestra — the player's name was Phasey — and Costa is reputed to have retained the services of two serpent players, who were to be seen with their obsolete instruments well up at the back of the choir among the men singers. His noisy array of wind instruments was notorious. Costa thought more of his three trombones than of his much-advertised sixteen double-basses; yet the performances were impressive, because Costa brought unanimity of purpose into the orchestra.

Let the worst be said of him first. Sir George Grove described Costa as 'a splendid drill-sergeant, he brought the London orchestras to an order unknown before. He acted up to his lights, was thoroughly efficient as far as he went, and was

THE PHILHARMONIC PERIOD

eminently safe', while Sir August Manns went further: 'His parts were full of cues, and it was more upon these than upon rehearsal that he relied for successful performance.' These remarks were made, it is true, by men whose lives were devoted to ideals that Costa tended to reject, but they were obliged to build on the foundations Costa had laid, and it is unfortunate that more emphasis has not been laid on Costa's early work with the Italian Opera and the Philharmonic Society.

The history of the orchestra is not only the history of conductors, leaders, and composers, but of an anonymous majority of players for whom the orchestra was a means to daily bread and little else besides. In 1871 the Rev. H. R. Haweis wrote a book with the naïve title *Music and Morals*, in the course of which he described the drawbacks of an orchestral player's life, how his enthusiasm was damped by

> The weary repetition of what he knows for the sake of other players who do not know their parts, the constant thwarting of the gifted players by the stolid ones, and the tension of long and harrowing rehearsals under conductors who do not know their own minds, or who cannot impart what they do know to the players, or who are so irritable, cantankerous, and at the same time so vexatiously exacting, as to destroy every particle of pleasure or sympathy in the breasts of the executants at the very moment when these qualities are most indispensable to the execution of the music. Then there is the cheerless wear and tear of regular orchestral life. The pantomime music, not in moderation and once in a way, but every night through a protracted season; for we are afraid to say how long the pantomime goes on after the departure of that inveterate bore, Old Father Christmas.
> Then really excellent players are occasionally subjected to the demoniac influences of that rhythmic purgatory known as the Quadrille Band; or the humbler violinists are to be met with, accompanied by a harp and a cornet-à-piston, making what is commonly understood to be music for the dancers in 'marble halls' or anywhere else, it matters little

enough to them. Shall we blame them if they look upon such work as mere mechanical grind — as the omnibus-horse looks upon his journey to the city and home again — a performance inevitable, indeed, but highly objectionable, and not to be borne save for the sake of the feed at the end? Then we must not forget the low salaries of many orchestral players, the small prospect of a slow rise, and the still smaller chance of becoming leaders in any orchestra worth leading. Or again, the weariness and disgust of your efficient men at seeing themselves kept out of their rightful places by old, incompetent players.
It is quite impossible to say at what age a man gets past his work, but the conductor of every orchestra knows very well who it is that mars the whole; and it is quite notorious that whatever inferiority there is in our leading orchestras, is chiefly owing to the fact that a conductor cannot very easily get rid of men who have grown infirm in their places, and who would have retired long ago from any foreign orchestra as a matter of course.

These were the conditions that tended to drag orchestral playing down to mediocrity. Not even Mendelssohn, with his personality and reputation, could entirely bend the players to his will or secure order. One had to compromise with them. Costa put an end to that. In him the men recognized their master: a stern one, but a just one. He set himself the same example of strict discipline. If Costa demanded punctuality he was always punctual himself; if he was insistent on attention to detail, he himself never relaxed; if it was certain that an erring player would be reprimanded, it was equally true that in all disputes with the management Costa acted as the leader of his men and insisted on their having a fair deal. By such means Costa succeeded in bringing discipline into the orchestra at the Italian Opera, of which he became conductor in 1833. The best contemporary authority on Italian Opera in England is H. F. Chorley, who in his *Thirty Years' Musical Recollections* (1862) says that:

> From the first evening when Signor Costa took up the

THE PHILHARMONIC PERIOD

baton — a young man from a country then despised by every musical pedant, a youth who came to England without flourish, announcement, or protection ... it was felt that in him were combined the materials of a great conductor: nerve to enforce discipline, readiness to second, and that certain influence which only a vigorous man could exercise over the disconnected folk who made up an orchestra in those days.

This was the first, and most generally recognized reform that Costa introduced to orchestral music in England. The second, less generally recognized, was his adoption of scientific principles in the disposition of his forces on the platform. A glance at the illustration from *Punch* on page 165 will show the general arrangement of an orchestra in England at that time; to clarify Costa's reform in this arrangement, a sketch of the placing of the instruments in the Hanover Square Rooms at Costa's first concert for the Philharmonic Society on March 16th, 1846, is given on page 175. Costa's arrangement is the basis of our modern disposition of forces, while Jullien's harks back to the eighteenth century, when the conductor sat at the harpsichord with his orchestra grouped round him. Although Jullien doubtlessly regarded himself as a revolutionary conductor, it will be seen that in essentials Costa was more advanced. The orchestra Costa employed was only seventy-eight strong, composed of 15 first violins, 14 second violins, 10 violas, 9 'cellos, 9 double basses, 2 each of flutes, oboes, clarinets, bassoons, and trumpets, 4 horns, 3 trombones, 1 ophicleide, and 1 drummer, but it was more sonorous than Jullien's vast crowd. The *Illustrated London News* reported that:

> A greater triumph never prevailed. The oldest members of the Society frankly admitted, that never before in this country had the great symphonies and overtures been so marvellously executed; and the critics have one and all handsomely acknowledged the genius of Costa in the management of his forces. It is necessary to enquire into the secret of such an important result, and the first

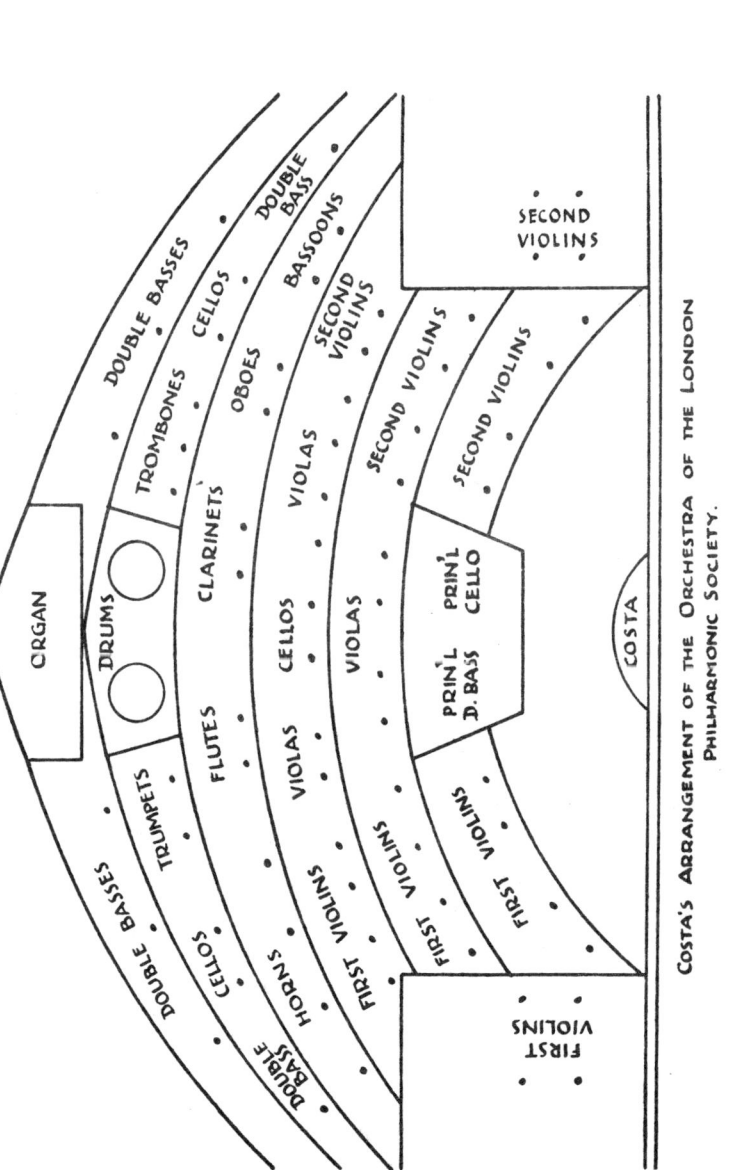

Costa's Arrangement of the Orchestra of the London Philharmonic Society.

THE PHILHARMONIC PERIOD

point to be considered is the disposition of the orchestra. To persons acquainted with the general arrangement of an orchestra, it will be seen that a complete revolution has been effected. The rapid and almost perpendicular rise of the old plan was long a matter of complaint — now, the elevation is reduced considerably; players with drums and trombones, etc., are no longer perched up in the roof to drown the stringed instruments, which are in a valley, with a formidable array of double basses in front, effectively to stifle the melody of the first violins. Costa has got rid of such monstrosities; and studying the principles of acoustics, he has sought, and successfully, to blend the various tones.

With Costa's control the title of 'leader' vanished from the Philharmonic programmes.[1] The necessity of principal players of the string groups was, however, recognized, and it will be seen that the positions of the principal 'cello and principal double bass are not unlike those occupied by their leaders in the Handel Commemorations of the eighteenth century.

For the first time Londoners heard an orchestra that was a single musical instrument instead of a body of individuals. Costa's personality dominated the performance, and although he never pretended to impose his own 'reading' of a score on his audience, the effect he produced was distinctive.

> ... the band began Haydn's Symphony in B flat, No. 9, at the conclusion of which the applause was deafening. Indeed, the sensation created by the colouring given to this most hackneyed work was one of utter amazement to the old Haydnites, who exclaimed loudly that they had only heard it for the first time that evening. Mr. Lockey was announced to sing *O cara, immagine,* from the *Zauberflöte,* but being indisposed, his place was taken at a few hours' notice by a Mr. Rafter, a pupil of Crevelli. It was Mr. Rafter's *debut* in London ... Here another innovation was remarked; all the violins played in the accompaniments,

[1] The term *chef d'attaque* was used by a contemporary journalist to describe the principal first violin.

COSTA AND BERLIOZ

instead of being taken off, as heretofore, and yet such a *piano* was preserved that the voice of the singer was fully sustained, and not drowned, as formerly.

Here, then, is the true Costa. The Costa of the mighty choirs at provincial festivals was a Costa playing on a quite different instrument, aiming for different effects from those of a Haydn symphony. Every lover of the orchestra knows how superior in taste is symphonic music to choral music, every orchestral player knows from experience that one is a grind and the other a stimulant: for years a story has gone round band-rooms of a 'cellist who dreamt he was playing in a *Messiah* concert, and awoke to find that he was; but oratorio was too much respected in Victorian times (and orchestral players ignored) for such a point of view to find general acceptance; Costa of the Philharmonic Society is forgotten while Costa of the festivals lives.

To some extent this was because Costa remained so short a time with the Philharmonic Society — from 1848 to 1854 — and also because of his personal character. Costa was an austere man, keeping himself aloof from his fellow musicians, and certainly not given to explaining his theories other than by the results he obtained in performance. Combined with this was an absolute honesty: a freedom from guile that sometimes alienated him from his colleagues, and generally surprised them. A story illustrating Costa's character is told by the music critic, Joseph Bennett, all the more remarkable because it happened almost at the end of Costa's long life.

Music critics in the nineteenth century were a mixed lot; not all of them as well-informed as they might have been, but all conscious of their own power. Only once did Joseph Bennett receive a visit from Costa, and his conscience was badly shaken by the event. Bennett had been asked to write an analysis of Costa's oratorio *Eli* for use at a Glasgow festival that Costa was to conduct. Bennett did so, but wrote in such a way that anyone who chose to read between the lines would see that the great idol Costa had feet of clay. Consider Bennett's consternation then, when he saw Costa's carriage stop before the house where he was staying in Glasgow for the term of the festival, and Costa alight and enter the house — a block of flats. Bennett naturally thought that Costa, the terrible drill-sergeant, was coming to 'have it out' with him. Slowly, the elderly conductor toiled up the stairs to the third floor, and his card was taken in to Bennett. He had not come to quarrel, however, but to thank Bennett for his analysis. Costa was not the sort of man who read between the lines, and Bennett's irony had passed unnoticed. 'I do not mind confessing that I was relieved by this courtesy, and more than a little sorry that the whole of it was not deserved', said Bennett. But the incident was not closed; when Costa had expressed his admiration and gratitude to Bennett for what he had done on behalf of *Eli*, Costa, with his old-time courtesy still unsatisfied, invited Bennett to accompany him to the hall in his carriage, where they parted, as it happened, finally, for Costa died shortly afterwards.

This dignified aloofness was an adjunct to Costa's greatness. It is necessary for a conductor to be able to shut himself off from others at times, for his work involves the study of scores (and Costa was always careful to edit his scores and parts before bringing any music to the notice of his audience), but in their performance he is a public figure and liable to be drawn into a great many public functions. Costa's aloofness was his protection from an over-enthusiastic public; a means by which he could escape into the retirement necessary for study.

It dominated his relations with the directors of the Philharmonic Society. When he became their conductor he did so on the understanding that he should not be asked to attend any meetings of the board, but should have a free hand with the orchestra. He made it clear, also, that he would only perform what he called 'worthy' music. To these terms the directors agreed,[1] and their confidence in Costa was rewarded by the advancement of the standard of their performances he immedi-

---

[1] Their financial position was strained prior to 1846, and their conductors lacked initiative. Bishop was more at home with the Ancient Concerts, and Moscheles failed to make much of new compositions.

ately brought about. On the business side of the Society's activities an important change of policy was their decision to issue tickets for single concerts, thereby opening the opportunity for all music lovers who could afford the fee to attend Philharmonic concerts. Previously admission had been reserved for an exclusive circle of Members, Associates, and their friends. This suggestion came from a business firm: the house of Cramer, Beale and Co., that, from a small beginning under J. B. Cramer, had developed into the leading music-shop in London.

Beyond this point it would appear that the advice of business houses was not acceptable to the directors of the Philharmonic Society. No doubt Costa could have attracted much larger audiences than the Hanover Square Rooms would hold, but the Philharmonic Society was interested in the artistic standard of performance of orchestral music — not in its popularization or exploitation for private profit. The policy of the Society continued under Costa much as before. Haydn, Beethoven, and Mendelssohn were the most respected names in their programmes, and Costa handed over the orchestra to Mendelssohn in 1847, when, on what proved to be his last visit, he conducted his *Scotch Symphony*[1] and *Midsummer Night's Dream* music. Costa's personal taste is shown in an increase of Italian works, notably by Rossini, and in a complete absence of the works of J. S. Bach. Bach was a composer practically unknown in England, but six of his works had been performed at Philharmonic concerts prior to 1846.

The pianist-composer coming into prominence in Costa's earlier Philharmonic programmes was Sterndale Bennett, who played his *Caprice* and *Pianoforte Concerto in F minor*, and whose overtures *The Naiades* and *Parisina* Costa conducted. Costa, however, does not appear to have regarded the compositions of Berlioz as 'worthy' music, and trouble arose because of this.

While it may be true that Berlioz made little impression in

[1] The Victorian title 'Scotch' has been retained only when referring to this symphony In all other cases the word Scottish is used.

THE PHILHARMONIC PERIOD

1849 under the wing of Jullien the charlatan, he earned the respect of British musicians at the Great Exhibition of 1851. With him on the jury which sat to consider the merits of the musical instruments exhibited was Dr. Henry Wylde, a member of the Philharmonic Society. Wylde broke away from the Philharmonic Society in 1852 and founded, in conjunction with Sir Charles Fox and others, the New Philharmonic Society. Exeter Hall was chosen as the meeting place and Berlioz as conductor. While the aims of the new society were much the same as those of the older one, special prominence is given in their prospectus to their intention to give modern and native composers a favourable opportunity of coming before the public. Clearly, in the view of the founder of the new society, the old Philharmonic Society was not carrying out its function satisfactorily. The New Philharmonic Society was, however, to be run for private profit, in which respect it differed from the old, so that the existence of two societies of the same type endeavouring to attract the same audience, but differing in their economic aims, may serve as an indication of the influence of motive on artistic developments.

The New Philharmonic Society started with an orchestra of ninety players, on a basis of sixty-eight strings subdivided into 16 first violins led by Sivori, 16 second violins led by Jansa, 12 violas led by Goffric, 12 'cellos led by Piatti, and 12 double basses led by Bottesini. It was therefore a larger orchestra than that of the original Philharmonic Society, and of its quality there can be no doubt. But in order for a profit-making society to succeed it is necessary that it should make profits: this the New Philharmonic never did. Exeter Hall was never filled, and so after their second season they moved to a smaller hall — St. Martin's — still without recuperating their lost finances; they then moved back to Exeter Hall and renounced their capitalism, advertised widely that the profits would be devoted to charities — those of the first concert of 1855 (if any) being earmarked for the Asylum for Idiots. There were no profits, and Wylde's programmes by this time had become quite un-

COSTA AND BERLIOZ

distinguished, while at the older Philharmonic Society Richard Wagner was that year a veritable storm centre.

So much for the virtues of capitalism in the 'fifties. But in their first season, before profit and loss had become a touchy subject in New Philharmonic circles, the society scored a decided triumph with Berlioz as conductor. The programme included Beethoven's little-known *Triple Concerto in C* for pianoforte, violin, and 'cello, which Silas, Sivori, and Piatti saved from its usual tiresomeness by some excellent playing. Bottesini also greatly impressed the audience with a solo on the double bass. This instrument has little appeal for us as a solo instrument, but Bottesini, and, before him, Dragonetti, were in great demand as soloists. Bottesini used a small instrument of the type called a chamber bass, strung with harp-strings, but Dragonetti used an exceptionally large bass, and besides his skill as a soloist he was noted for his partnership with Lindley the 'cellist at the opera and elsewhere; these two players accompanied all dry recitatives on their instruments — elsewhere a pianoforte had to be used.

London was thrilled with Berlioz's masterly control of the orchestra. 'Berlioz as a conductor must be placed in the first rank of orchestral generals', wrote a critic with an original gift in metaphor. 'The rush of the violins in *Anacreon* at the close quite electrified the assemblage.' At the same concert — the second of the series — Dr. Wylde killed his own *Pianoforte Concerto in F minor* by his inexperienced conducting. He would have done far better by leaving the baton with Berlioz, but it was one of the tragedies of this Society that Dr. Wylde never seems to have realized his own limitations. This was the man who was opposing the great Costa!

Berlioz gave two performances of his own *Romeo and Juliet*, the first of which was much discussed as a novelty but not disliked. At its second performance the audience accepted it as a masterpiece of orchestration, and all agreed that its execution under the composer himself was admirable. Those critics who had denounced Berlioz's compositions before they had heard them found that public opinion was against them. 'He has won the suffrages of our musical audiences by the magical influence of his genius: it has been a battle over personal prejudices, intense intolerance, artistic ignorance, and bigotry; but the victory has been for art development and progress against the standstill purists and dogmatists.' It does not read like an expert opinion, but it is proof of Berlioz's popularity.

If expert opinion is wanted, it is provided by the Philharmonic Society's programmes for the following year. At their sixth concert Berlioz was invited to conduct a selection of his works. He chose *Harold in Italy* (Sainton played the solo viola part), the overture *Le Carnaval Romain*, and an air, *The Repose of the Holy Family*, from the oratorio *The Flight into Egypt*. Berlioz's opinion of the Philharmonic Society's orchestra is evident in his remark: 'One rehearsal will be ample with your orchestra.'

Costa, of course, had to agree to Berlioz's appearing at the Philharmonic Society's concert. He accepted the suggestion with his usual dignity and courtesy, but it was a breach of the conditions under which he had taken office. The suggestion that Berlioz should appear did not originate from Costa. Throughout 1854 Costa remained with the Society, but announced that the strain of his numerous engagements would necessitate his resignation at the end of that season. The directors were naturally unwilling to lose Costa's services, and continued to hope that Costa would change his mind right into January of 1855. It was obvious that Costa's health was not affected by his many engagements, for he continued to direct these with as much vigour as before. He remained, however, courteous but inflexible. Finally, the directors, on the eve of their new season, invited Richard Wagner to take Costa's place, and Wagner accepted.

WAGNERIAN INTERLUDE

THE nineteenth century was an age of belligerent individualists. Men were proud to say that they followed an unswerving course, guided by their own opinions, and were prepared to defend those opinions against all comers with every resource at their command. Not least among these belligerent individualists was Richard Wagner, who, not content with political revolution in Saxony, must needs follow up with a tract on *Das Judenthum in der Musik*, in which the music of Jewish composers was held to be a sham, without life or direction. It was based on the theory that the Jews were not a nation, but merely cuckoos in the nests of other nations, and most of Wagner's conclusions are sealed with examples from the music of Meyerbeer and Mendelssohn. It follows, then, that Wagner would have enemies wherever the name of Mendelssohn was respected, and nowhere was this name more respected than in London.

Nevertheless, Wagner was invited to succeed Costa as conductor. In their desperate efforts to persuade Costa to change his mind, the directors of the Philharmonic Society had left the appointment of another conductor dangerously late. Spohr could not leave Cassel, Berlioz was not available. Sterndale Bennett was in London, but too busy with his Bach Society and his numerous pupils to take on another post at short notice, but Wagner was languishing in Zurich with no public duties to keep him there. A messenger was sent and the appointment made on the spot.

In London the critics prepared to attack. They were just as belligerent as Wagner, and as proud of their individuality. Chief among them was J. W. Davison, of *The Times*, as ardent an individualist as could be imagined. 'Where his personal likes and dislikes were not concerned', wrote Sir Frank Burnand, 'his criticisms were reliable; but where there was bias, then to read between the lines was an absolute necessity in order to get

THE PHILHARMONIC PERIOD

at anything like the truth.'[1] Davison openly announced his intention not to give Wagner a chance,[2] and Davison controlled pens other than his own. He was editor-in-chief of the *Musical World*, a paper owned by his brother William, and employed as assistant editor one Desmond Ryan, who was also music critic to the *Standard*. Lest the disclosure of these ramifications should suggest that the bulk of the London press was in Davison's pocket, let it be known at once that there was no more stubborn antagonist to Davison than C. L. Grüneison, of the *Morning Post*, who attacked Wagner too, and so did H. F. Chorley of *The Athenaeum*, a man as exclusive and difficult to approach as Costa was, and with strong ideas on the nature of opera. Chorley had a gift of concentrated invective that made him not a few enemies, but for once he was not attacking alone. He regarded *Tannhäuser* as an insult to his intelligence.

Against these leaders of opinion must be placed a paper that avoided controversy, and was content to report those things that most interested its readers. The *Illustrated London News* catered for the Victorian family, with pictures for those who disliked argument. 'Our business', said its founder, 'will not be with the strife of party, but with ... the home life of the empire, with the household gods of the English people, and above all of the English poor.' This ship carried no big guns, cruised only in calm waters, and showed its passengers the things they wanted to see.

On March 5th, 1855, Wagner arrived in London and established a *causa belli* by his neglect of a certain custom on which critics in those days set some importance — that of calling upon the critics to introduce himself and solicit their good opinion. It was a foolish custom, laying the critics open to suspicion of taking bribes, but it was a slight to ignore it. Wagner, however, did not offer the easy target he might have done, for his first concert contained only such pieces as were well known to the Philharmonic Society's audiences. The symphonies were by

[1] Sir Frank Burnand, *Records and Reminiscences*.
[2] Dr. F. C. W. Praeger, *Wagner as I knew him*.

WAGNERIAN INTERLUDE

Haydn and Beethoven (*The Eroica*), with Mozart's *Magic Flute* overture, Mendelssohn's *Fingal's Cave*, and Spohr's *Dramatic Concerto* for the violin, played by Ernst. 'His appearance at the head of the Philharmonic band enabled the public to judge only of one thing — his capacity as a *chef d'orchestre*: a point which that one evening settled beyond all question. Though the whole orchestra — till the rehearsal two days before — were utter strangers to him, yet that single rehearsal had established so thorough an understanding between them, that, at the concert, every piece was performed with a clearness, spirit, and delicacy that has never been surpassed; and this was the more remarkable as his manner of marking the time, and his readings of many passages, differed materially from those of his predecessor.'

Wagner, in fact, laid down his baton in straightforward passages and resumed control of the orchestra when a change was needed, or if the orchestra showed signs of faltering. The players were satisfied with him, and if he was, as people said, a 'Marat of music', come to England to guillotine the classical composers and establish a republic of noise, he was, they thought, going about it in a somewhat strange way.

Wagner himself has left on record his opinion of the Philharmonic orchestra and the policy of its directors; it proves that he had a great respect for them though he was by no means unaware of their failings. His criticism is what we should expect from a good conductor finding himself in charge of a Costa-trained orchestra.

> A magnificent orchestra, as far as the principal members go. Superb tone — the leaders had the finest instruments I have ever listened to — strong *esprit de corps* — but no distinct style.

That is sound and not unsympathetic criticism. His other remarks are less favourable, but patently true:

> The fact is that the Philharmonic people — orchestra and audience — consumed more music than they could digest.

THE PHILHARMONIC PERIOD

As a rule an hour's music takes several hours' rehearsal — how can any conductor, with a few hours in the morning at his disposal, be supposed to do justice to monster programmes such as the directors put before me?

Two symphonies, two overtures, a concerto and two or three vocal pieces at every concert! The directors continually referred me to what they called Mendelssohnian traditions, but I suspect that Mendelssohn simply acquiesced in the traditional ways of the Society.

One morning, when we started a rehearsal of the *Leonora* overture, I was astonished, for everything appeared dull, slovenly, inaccurate, as though the players had not slept for a week. Was this to be tolerated from the famous Philharmonic Orchestra? I stopped and addressed them in French, saying that I knew what they were capable of and I expected them to do it. Some understood me and translated to the others; they were taken aback, but knew that I was in the right. So we began again and the rehearsal finished off quite well.

This must have been at the sixth concert, when Wagner had been conducting the Philharmonic orchestra for nearly three months. By that time Wagner had had time to try out *Lohengrin* and *Tannhäuser* on the London public, and to take the measure of the critics. *Lohengrin* came first — the *Introduction, Bridal Procession, Wedding Music,* and *Epithalamium.* Nothing that has not since become popular. Yet these excerpts gave the antagonistic critics the chance for which they had been waiting. The *Musical World* said:

> We hold that Richard Wagner is not a musician at all, Look at *Lohengrin* — that *best* piece; it is poison, rank poison. All we can make out of *Lohengrin* is an incoherent mass of rubbish with no more pretension to be called music than the jangling and clashing of gongs, and other uneuphonious instruments.

while the *Sunday Times* was equally vindictive:

> Richard Wagner is a desperate charlatan — scarcely the most ordinary ballad-writer but would shame him in the

WAGNERIAN INTERLUDE

creation of melody and no English harmonist of more than one year's growth could be found sufficiently without ears and education to pen such vile things.

These are examples of criticism that have been freely quoted in books on the Philharmonic Society. They are quoted because they are the views of known authorities. How far history can lie as a result of quoting prejudiced authorities may be seen from this report of the same concert, taken from the *Illustrated London News*:

> Wagner, though in great vogue in Germany, has been hitherto quite unknown in this country: and this specimen of his powers as a composer excited great curiosity and interest. The public has been told that he is a musical revolutionist, whose object is the destruction of existing greatness — who seeks that he may raise himself to supremacy in their room. Such, we are informed is the purpose oɩ his critical writings; and, it is added, his extravagant doctrines are illustrated by equally extravagant compositions. It was with no small surprise, therefore, that the public, thus prepossessed, listened to Wagner's music on Monday evening. In place of finding it to be obscure, unintelligible, and studiedly unlike anything ever heard before, they discovered that it was clear, simple, melodious and not at all hard either to perform or to comprehend. The audience were delighted: their prejudices were overcome by their feelings, and they applauded frankly and warmly; all but the professional 'native talent' clique, who comforted themselves by trying to convince everybody who would listen to them that the music was conventional and commonplace ... With regard to Wagner's character as an orchestral conductor, there was not, on this occasion, a single dissenting voice. His great skill, and its happy results, were felt and acknowledged from beginning to end of the concert.

At the same concert Wagner conducted Beethoven's *Ninth Symphony*, and wrote a penetrating if rather verbose analysis of the work. Undoubtedly Wagner was doing his best to serve the society, and he was used to opposition. He was not a free agent in his choice of programmes, as Costa had been, and the directors continued the policy of including chamber-music works in their programmes, but very rarely, and only once during Wagner's year of office, when Spohr's *Nonett* was given. There was, however, a fair amount of Mendelssohn performed that year — the second and third Symphonies, the *Violin Concerto*, the *Midsummer Night's Dream* and *Hebrides* overtures, which Wagner had to conduct. He did so conscientiously (and indeed, even in *Das Judenthum in der Musik*, he gives Mendelssohn credit for knowing his job), but saw no reason why his personal dislike of Mendelssohn should be concealed. Wagner ostentatiously removed his gloves, and threw them on the floor after handling a Mendelssohn score. So much for the Mendelssohn tradition of the Philharmonic Society.

With some of the patience that their founders had shown to Beethoven, the directors of the Philharmonic Society supported Wagner, but rather, one feels, from a sense of duty than from any love of his music. The overture to *Tannhäuser* was given at the fifth concert of the season, but by this time the enthusiasm of the general public was beginning to cool. The *Illustrated London News* no longer felt it advisable to applaud the work as it had applauded *Lohengrin*, but could not join the prejudiced. The *Illustrated London News* therefore sat on the fence:

> The overture to Wagner's much talked of opera *Tannhäuser*... was most carefully executed and listened to with much curiosity and interest. Opinions were much divided with respect to its merits. Some deemed it, though wild and eccentric, a work of originality and genius, while others condemned it *in toto*.

The way was open for Wagner's antagonists. 'Why, this is Brummagen Berlioz!' said Sterndale Bennett in the hall, and H. F. Chorley wrote:

> The overture to *Tannhäuser* is one of the most curious pieces of patchwork ever passed off by self-delusion for a com-

WAGNERIAN INTERLUDE

plete and significant creation. The instrumentation is ill-balanced, ineffective, thin and noisy.

It was at a rehearsal for the concert following this that Wagner had to reprove the members of the orchestra for slackness. How far the antagonism would have spread but for the Queen and Prince Consort there is no knowing, but the seventh concert of the season was a command performance at which the *Tannhäuser* overture was repeated. Their majesties spoke with Herr Wagner for a long time in his own language, and since they had been responsible for the choice of programme, and had, moreover, appeased the native talenters by including Macfarren's *Chevy Chase* overture, the Philharmonic Society retained that decorum for which it had always been noted. Wagner is reported to have said that *Tannhäuser* did not worthily present him to Royalty, but it was his last chance, for none of his works appeared on the programme of the Society's eighth and last concert of the season, although Mendelssohn and Meyerbeer were both represented. The influence of Prince Albert on British commercial policy and on Victorian social life were no doubt considerable, but he was never popular, and in certain quarters was regarded with a most uneasy suspicion. He liked the role of *Deus ex machina*, and believed ardently in the beauty of all things German. But he was a cultured man, a humane man, a sincere man, and for these reasons the actions of his opponents too often appear petty.

SIR CHARLES HALLÉ

APART from the actual work of the Philharmonic Society in providing high-class orchestral concerts in London, there were many developments in musical life that owed their initiation to members of that Society. The most important of these is undoubtedly the founding of the Royal Academy of Music in 1822. This was first suggested by T. F. Walmisley at a meeting of the Society held on April 13th, 1822, but was handed over on July 20th to an enthusiastic group of aristocratic amateurs, led by Lord Burghersh. Most of the original staff of the Academy were chosen from members of the Philharmonic Society. From a more modest beginning in the middle of the century came another innovation, when in 1845 John Ella, a violinist member of the Society, started his Musical Union for the performance of chamber music before smaller audiences than frequented the Philharmonic concerts. This venture was meeting with some success when Costa joined the Philharmonic Society in 1848, though it was not until 1859, when the Monday Popular Concerts in St. James's Hall superseded the Musical Union, that the circumscribed definition of chamber music now in use came to be generally accepted. At first Ella found that pianoforte solos played by Arabella Goddard and songs sung by Sims Reeves were most popular with his patrons, and he had to get a living. Nevertheless, the Musical Union had its place in the reform of Victorian ideas of fashion in music, not so much through Ella, perhaps, as his artists.

The Musical Union was a source of amusement to contemporary critics because of Ella's pronounced respect for class distinctions. For many years Ella had directed private concerts for Lord Saltoun, Sir George Warrender, and similar patrons of music among the nobility, and his Musical Union was designed to reproduce such concerts in a more public way, much as John Banister in the seventeenth century had capitalized the

SIR CHARLES HALLÉ

idea of chamber-music performances which were a feature of contemporary social life among musical amateurs. Ella placed his performers in the middle of the room, and it was said that the first ring of chairs round the performers was occupied by duchesses, other ranks of society being accommodated according to order of precedence.[1] Such an audience in the eighteenth century might have had musical taste of a high standard, but this was by no means to be relied on in Ella's day; Ella met the situation by restricting his programmes to three classical works at each concert, and by introducing annotated programmes. Popular artists with serious musical ideals, like Sir Charles Hallé, did the rest.

Hallé was a fine raconteur, and loved to tell stories of his audiences, one of which concerned an engagement to play at a fashionable London party. Society was there in full force, and the rattle of tongues and teacups continuous, so that it was some time before the hostess was able to take advantage of a lull to ask Hallé to play. As soon as she did so the chatter recommenced, but Hallé finished his piece and was duly congratulated on a fine performance. Later he was called upon again, with similar results, except that the hostess declared that she liked the second piece better than the first. After a third solo had been given with the guests' tongues, if anything, even more in evidence, the hostess became quite effusive in her admiration of how he had begun with the good, proceeded to the better, and finished with the best.

'But', said Hallé, 'nobody listened.'

'Sir Charles!'

'Not even your ladyship, or you would have noticed that I played the same piece each time.'

Despite such behaviour, Hallé made rapid progress in favour among the wealthier London patrons of music, but he never allowed his artistic ideals to be governed by them. Soon after coming to London in 1848 he appeared at one of Ella's concerts

---

[1] Not always. It sometimes happened that a duchess gave her ticket to her maid.

for the Musical Union, and expressed a wish to play one of Beethoven's sonatas. Ella tried to dissuade him, saying, 'They are not works to be played in public'. But Hallé would not be put off; he played the *Sonata Op. 31, No. 3 in E flat*, and the result was that a craze swept fashionable London; afternoon parties were arranged to hear this sonata, and Hallé's pockets grew heavier in consequence. Later he broke the Hummel concerto tradition of the Philharmonic Society, for at his first appearance on their platform he played Mendelssohn's *Concerto in D minor*, again without offence, for most of the cherished traditions were seen to be not worth fighting for when resolutely attacked.

Hallé's pianoforte playing had much in common with Costa's conducting. It was an honest reproduction of the score, in no way consciously twisted to suit the performer's personality. It was a form of honesty on the decline among soloists, and doomed by the end of the century to be actually unpopular. Hallé is said to have been a cold pianist, and Costa an unimaginative conductor. It is not possible after so long a time to judge impartially of the merits of these two men as interpreters. It is not, indeed, as an interpreter that Hallé should be judged, for judgment of interpretations involves comparisons, and Hallé spent his main efforts introducing great orchestral music to audiences in Manchester, where it had rarely been heard before.

Manchester's importance had increased greatly as a result of the expansion of the Lancashire cotton trade, and all the fruits of the Industrial Revolution could be observed ripening or rotting in the city's life. Hallé received his invitation to live in Manchester from a group of German merchants who resided there. Within the limitations of their environment these merchants strove to enjoy the cultural pleasures to which they had been accustomed in Germany; this cultural life centred round their *liedertafel*, a club which, as its name implies, was originally a singing-party but which was really more than this. Hallé was invited to come to Manchester in the hope that he would be able to improve the standard of instrumental music in the city, which, from Hallé's own description, seems to have been badly behind the times:

> Not long after my arrival in Manchester I had occasion to hear one of the oldest and most important musical societies of the town, called 'the Gentlemen's Concerts' from the fact that it was originally founded in 1774, I believe, by amateurs, twenty-six in number, who constituted what may be called the orchestra, but who all and every one of them played the German flute! In course of time other instruments were added, and in 1848 the modern orchestra had been completed for more than a score or two of years. The society was wealthy, would-be subscribers having generally to wait three years before room could be made for them; in consequence every artist of renown who had visited England had been engaged, and the older programmes of the concerts were remarkably rich in celebrated names. At the concert which I attended Grisi, Mario, and Lablache sang; but the orchestra! oh, the orchestra! I was fresh from the *Concerts du Conservatoire*, from Hector Berlioz's orchestra, and I seriously thought of packing up and leaving Manchester, so that I might not have to endure a second of these wretched performances. But when I hinted at this my friends gave me to understand that I was expected to change all this — to accomplish a revolution, in fact — and begged me to have a little patience.

Hallé stayed and was invited to play a concerto at the next concert. He chose Beethoven's *Concerto in E flat,* and Zeugher Herrmann was invited to come over from Liverpool to conduct the orchestra, which Hallé says he did with great skill, accomplishing all that could be accomplished with the unsatisfactory players under his command.

This was in August 1848. In the same month Chopin came, but his playing was little understood in Manchester and he moved on to Scotland. This, then, must be taken as the standard of musical taste among the well-to-do of Manchester when

THE PHILHARMONIC PERIOD

Charles Hallé arrived there. The anticipated invitation to become conductor of the Gentlemen's Concerts came the following year, 1849, and Hallé accepted the post only on condition that the whole of the orchestra should be dismissed and its reorganization left entirely in his hands. He succeeded in attracting to Manchester a number of good players from London, whom he used to replace the less competent of the Manchester men, and set about the work of reorganization. The arrangement of the players on the platform was his first consideration; the old arrangement, with the double basses in front, gave way to a more rational disposition of forces, and the results were approved by the subscribers. Manchester's place in the history of the orchestra in England may be said to date from this event.

Hallé's scheme was not complete with the reorganization of the orchestra, however; he wanted a choir as well. In this he had none of the difficulty that had beset his instrumental scheme, for Lancashire voices were of good quality, and Manchester had an educational tradition in choral singing lately improved by Mainzer. Hallé had no knowledge of Manchester choral traditions; his ideas of choral training came from his childhood recollections of the German *Gesangverein*, in imitation of which he formed the Manchester St. Cecilia Society. This society he recruited from people in the best Manchester social circles — for, like the Gentlemen's Concerts, it was to be a select organization — and they met weekly under Hallé's direction for the study of good choral music. They were very keen, and both they and Hallé found much pleasure in these weekly meetings. From a small choir of fifty members in 1850 the Cecilia Society grew in membership, in influence, and, in Hallé's own words, 'contributed not a little to spread that intelligent love of the art which distinguished Manchester'.

Both the Cecilia Society and the Gentlemen's Concerts, however, were select — they affected only a small portion of Manchester's population. There were many music-lovers in lower walks of life whose activities in choral and brass band

SIR CHARLES HALLÉ

music played an important part in Lancashire life, but the well-to-do had no practical interest in these activities. The constant friction between employers and workers in the textile trade needed a special social technique if religious and intellectual life was to go on in nineteenth-century Lancashire, and a majority in all classes was determined that it should. There was not the same misunderstanding of Chartism in Manchester that there was in London: feeling ran higher, but the antagonists understood each other. The prospect of the crowds that would invade London for the 1851 Exhibition alarmed Londoners, but there was no such alarm when the 1856 Art Treasures Exhibition was contemplated in Manchester. Manchester was used to such things. Hallé and his friends were anxious only that music should be adequately represented at the exhibition, and it was felt that this could not be done with the orchestra Hallé had then at his command.

Ample means were placed by the committee at Hallé's disposal. With them he was able to attract players from London, Paris, Germany, Holland, Belgium, and Italy, who, with the best of his Manchester players, he formed into a first-class orchestra. This orchestra played every afternoon at the Art Treasures Exhibition, though Hallé himself conducted only on Thursdays. Thousands of people from the northern counties heard a symphony for the first time at these concerts, and Hallé watched with interest how the appreciation of such works grew keener as the season went on.

An end to the exhibition obviously had to come, and Hallé found himself in October 1857 faced with the painful prospect of disbanding the orchestra that had brought both pleasure and honour to Manchester. To prevent this he determined to engage the whole orchestra at his own risk for a series of weekly concerts during the autumn and winter seasons, trusting in the newly-awakened taste for symphonic music to reward him for his enterprise. The preparations took longer than he had anticipated, however, and the first concert did not take place until January 30th, 1858. It was poorly attended.

THE PHILHARMONIC PERIOD

Hallé was not disheartened. He reasoned rightly that the crowds that had thronged the exhibition did not come specially for the music, and that concerts offering nothing but music, and at necessarily higher prices of admission, could not expect the same popularity. It was a case for attempting some musical education for the whole community, and he resolved on a season of no less than thirty concerts to achieve this end. It must be borne in mind that this venture was a novelty to Manchester: the Gentlemen's Concerts had been exclusively for subscribers, no tickets were ever sold, and until Hallé requested it in 1850 they had never even published their programmes. (Hallé, indeed, had to press for this reform because he had strong objections to conducting concerts of such a clandestine nature.) His friends therefore regarded with dismay Hallé's decision to risk thirty public orchestral concerts, and it seemed at first that they were right, for the loss on the first concert was heavy, and was followed by similar losses week after week until Hallé's friends began to think how some means might be found of restraining him from his rash scheme. Not until the season had run half its course did any change for the better appear. The audiences gradually grew larger and more appreciative, until at last full houses succeeded each other. On the day after the thirtieth concert the Forsyth brothers (Hallé's agents) brought him a balance sheet and ten brand-new threepenny pieces — the profits. Perfectly satisfied with the artistic results — and a return of one penny per concert — Hallé at once made arrangements for a second series to be given during the winter season of 1858-59. From that time onwards the Hallé Orchestra has been the pride of musical Manchester.

In many ways the Hallé Orchestra has throughout its long history been at variance with generally accepted theories of how good music may be best introduced to a large audience drawn from all classes. Sir Charles Hallé never resorted to clowning as Jullien did — indeed, he had opposition in Manchester from a certain Mr. de Jong, who gave monster programmes in the Free Trade Hall at which arrangements by

SIR CHARLES HALLÉ

Jullien and Manns were played. At these concerts F. H. Cowen appeared as a solo pianist. Hallé had to keep down his costs, but he did this by keeping down the expense of too many soloists. His leader, Ludwig Strauss, frequently appeared as the violin soloist of the evening, and Hallé himself as pianoforte soloist (he played the Grieg Concerto as early as 1876). The Hallé Orchestra in the 'seventies comprised eighty players and was in every way a modern orchestra except that Hallé retained an ophicleide. In Manchester one could hear, for prices ranging from 7s. 6d. to 1s., all the best classical works, together with those of new composers like Brahms, Wagner, Gade, and Berlioz, of whom Hallé was very fond. For choral works a choir and orchestra of 350 was used.

It cannot be said that the work of the Hallé Orchestra has ever been easy. It has always had to supply the needs of some thirty large towns in the industrial north. Sir Charles Hallé was at one time even drawn into conducting local concerts in Liverpool, a city that has quite rightly always striven to develop an individual musical life in keeping with its own civic traditions. The importance of Hallé's work, taking place as it did in the midst of an area of exceptional pride in choral music, cannot be over-estimated, but it has taken many years for orchestral music to come into its own in provincial districts other than Manchester. Hallé himself had to do pioneer work all his life, but it would appear that only in London did he encounter opposition from the enlightened. To some extent this was due to metropolitan prejudice against provincial festivals. The most famous of these were gradually coming to realize the necessity of a good conductor: Sir Michael Costa became a familiar figure of the Birmingham, Glasgow, and Leeds festivals, Sir Julius Benedict conducted the Norwich and Sir Charles Hallé the Bristol festivals, but others, notably the Three Choirs festivals, retained their local conductors. It was not unusual for London orchestral players to have to pull the local conductors through these ordeals, and it may be that Hallé was not as careful before an uncritical audience as he was

THE PHILHARMONIC PERIOD

before his Manchester public. Certainly he had the reputation of over-working his players; he often took on contracts that he could only with difficulty complete. One such meeting took place at Bristol where the festival was to end with a performance of *Elijah* — a work of some length, and liable to become unduly long if a Victorian audience's demands for encores were to be satisfied. All would have been well had not Hallé accepted an engagement for an orchestra to play at a northern town on the day after. This meant that he and his players would have to travel all night in order to be at rehearsal on the following morning, and the only train available left Bristol just about the time his *Elijah* performance would normally be ending. In order to catch this train Hallé had to speed up the tempos of *Elijah*, arrange for vans to be waiting outside the hall for his orchestra's equipment, load up and leave Bristol while the audience was still gasping for breath. He did it, but such things are not good for a conductor's artistic reputation.

The wide area served by the Hallé Orchestra undoubtedly increased the strain that its members had to bear, but this could hardly be avoided. Hallé had drawn his players from a wide variety of countries, and he could only keep them in Manchester by the provision of regular employment.[1] The policy of casual engagement of orchestral players that had for so long been the rule in London could not be carried out in Manchester, without periods of unemployment for all the players such as would render their livelihood precarious. So Charles Hallé kept his men at work; he kept them too in constant rehearsal, and that is something that the London system could not do. If he sometimes fell from his best through over-working his men, at most such lapses were temporary and understandable.

Less excusable was his failure to do justice to Berlioz's *Faust* in London in 1880. There had been London performances of selections from this work, but Hallé was the first to perform it

[1] They were engaged for a six months' winter season, but most of them also obtained summer contracts at holiday resorts.

complete in the metropolis. Had Hallé taken the whole of his Manchester orchestra to London for the event, an outstanding performance would have been assured, but Hallé allowed the London agent, Chappell, to persuade him to cut expenses by using only a few Manchester players and make up the orchestra with London players; in fact, to acquiesce to the casual system general outside Manchester. Chappell made a profit at the expense of Hallé's reputation, for although Hallé visited London with his orchestra at later dates, he never succeeded in establishing there the reputation for orchestral finesse that he and his men enjoyed in Manchester. His last three London seasons were so ill-supported that he had to abandon them, and after 1891 London saw him and his orchestra no more.

Yet he had shown that the way to a really good orchestral style depended on having the orchestra regularly playing together, as they did on the Continent. England had for too long ignored this common-sense plan, and when the resolve came for its adoption they found that for too long they had been sowing their dragon's teeth. The abuses of the deputy system had taken firm root. But the uprooting of these is a later story.

BOOK THREE
THE NATIONALIST PERIOD

MANNS AND
GROVE AT THE CRYSTAL PALACE

WHILE Hallé was engaged in the creation of a first-class orchestra in Manchester, and overcoming the very real difficulty of stimulating a taste for orchestral music in many northern towns where local traditions favoured choral music, a similar movement was going ahead in London, starting, as Hallé's did, from an exhibition, and relying for support not on an exclusive body of subscribers but on the widest public obtainable. The Crystal Palace concerts show the gradual rise of a catholic taste over the prejudices of critics like Davison.

Davison's loyalty to the Philharmonic Society is shown in an article by him in *The Times* dated July 17th, 1862. It was the Jubilee year of the Society and Davison pointed out that

> Since its institution in 1813, the Philharmonic Society has, to use a homely phrase, seen various 'ups and downs'. Nevertheless, even in its darkest and most threatening periods, it has never once departed from the high standard which it set itself from the beginning, never once, by lowering that standard, endeavoured pusillanimously to minister to a taste less scrupulous and refined than that to which it made its first appeal, and to which it is indebted for a world-wide celebrity. Thus it has never forfeited the good opinion of those who actually constitute the tribunal which in this country adjudges the real position of the musical art, and who have invariably rallied round the 'Philharmonic' in its moments of temporary trial. Amid all kinds of well-intended, however bigotted, opposition, the Society has submitted to reform after reform, and preserved its moral equilibrium, a sign that its constitution is of the strongest and healthiest.

It is true that by 1862 some changes had appeared in Philharmonic concerts. They had become shorter, only one

THE NATIONALIST PERIOD

symphony and one concerto appearing on the same programme, and the name of J. S. Bach, which the great Costa had avoided, began to appear with reasonable frequency under his successor Sterndale Bennett. But Sterndale Bennett could not be regarded as a progressive musician in the latter half of the nineteenth century — he was chosen as conductor of the Philharmonic Society because of his loyalty to the Society's early traditions, from which, as Davison says, it had never departed. Sterndale Bennett had been a student at the Royal Academy of Music in its early years, when this institution was staffed by Philharmonic celebrities: Spagnoletti and Oury, W. H. Holmes, Thomas Attwood, Dr. Crotch, and Cipriani Potter had been his teachers, and Mendelssohn his friend. Loyalty to these he owed and gladly gave, but one would have thought he also owed some small loyalty to Schumann, whose tribute to Sterndale Bennett in the *Allgemeine Musikalische Zeitung* is generous in the extreme. Bennett was tardy in performing any music of Schumann's at Philharmonic concerts, however, and with no good reason, for the public was becoming interested in this modern composer. The first suggestion (from Mendelssohn in 1844) that the Philharmonic Society should perform Schumann's *Symphony in C* was not entertained, but there were performances of an overture in 1853 and the *B flat Symphony* in 1854 under Costa. In 1856, it is true, Sterndale Bennett conducted *Paradise and the Peri* at the last concert of the Philharmonic Society's season, but this was by Royal command, and it is singular that although the audience on this occasion was so numerous that the seating had to be rearranged to accommodate them, Sterndale Bennett gave no more performances of Schumann's works until 1862, when he chose the overture to *Genoveva*: for a Schumann symphony the members had to wait until 1864. It is significant that when Madame Clara Schumann appeared as solo pianist at two Philharmonic concerts in 1856 she played Beethoven and Mendelssohn, but at an afternoon pianoforte recital given in the Hanover Square Rooms, she played, besides some of her husband's music, Stern-

MANNS AND GROVE AT CRYSTAL PALACE

dale Bennett's *Two Diversions, Op. 17*, and *Suite de Pieces, Op. 24, No. 1*. To friends of Schumann, and those who wanted to hear new music, it appeared that modern composers were being slighted by Bennett, and on that belief the controllers of organizations giving other concerts were able to rise and establish a reputation for advancing the cause of music in a backward city. August Manns got the post of conductor at the Crystal Palace by reason of his willingness to introduce unknown compositions.

> 'I am much obliged to you for sending me the programmes', wrote George Grove to Manns in reply to the latter's application for engagement; 'I like them very much, and would give a great deal to have such music done at the Crystal Palace. The overture Op. 124 has only once been done in London in my recollection. Weber's Clarinet Concerto *never*. Berlioz's *Invitation à la Valse, never*. Nor do I ever remember hearing of Mozart having written a finale to Gluck's *Iphigenia*. Your playing of these rare works does you the highest honour.'

The fact that Weber's Clarinet Concerto had been performed at a Philharmonic concert in 1836 hardly affects the situation, for George Grove was then only sixteen years of age. Other keen music-lovers must have been equally unaware of this composition. Even more astonishing is a statement by Manns concerning Schubert:[1]

> I have reason to believe that my performance of the C major Symphony in 1856 was the first in England, although I remember hearing one of the members of my then very small band speak of a rehearsal of it under the late Dr. Wylde,[2] when, at the close of the first movement, the principal horn called out to one of the violins, 'Tom, have you been able to discover a tune yet?' 'I have *not!*' was

---
[1] Written in 1896.
[2] Dr. Wylde, conductor of the New Philharmonic Society, died in 1890. This paragraph needs some elucidation, for Wylde's band was not Costa's band, though they may have shared some players.

THE NATIONALIST PERIOD

Tom's reply. I quote these remarks made by two of the foremost artists in Costa's band (then the only band in England) in order to show how great the prejudice at that time against any compositions that did not come from the sanctified Haydn, Mozart, Beethoven and Mendelssohn.

Manns himself was not without prejudice, but his seeming slight to the Philharmonic Society by declaring that Costa's orchestra was the only one in London is not inconsistent with the facts. The Philharmonic Society drew the best players in London together for eight concerts each year, but it did not provide them with regular employment. It was possible for the Philharmonic Society to engage Costa's players from the Royal Italian Opera because Philharmonic concerts were traditionally held on Mondays and there was no opera on that day until 1860. The popularity of railway travel, however, brought more and more visitors to London each year during the London season, and it became possible to extend the opera performances to Mondays. A clause was therefore inserted in the contracts of Costa's players requiring their services on that day. Thus about half the players usually employed by the Philharmonic Society declined to accept engagements that year. George Hogarth was secretary of the Philharmonic Society at that time, and had the duty of conveying the feelings of the directors to the players concerned. This letter he reproduced in full in his history of the Philharmonic Society:

> Dear Sir,
> I have laid before the Directors of the Philharmonic Society your letter in which, for the reasons therein assigned, you decline to accept the engagement offered you in the orchestra next season, unless under reservations which the Directors find it impossible to accede. The number and dates of the concerts were expressly fixed by the last general meeting of the Society, and you are aware that the engagement with the orchestra must necessarily provide for their regular and personal attendance at every concert and rehearsal.

It is with sincere and deep regret that the Directors find themselves deprived of the assistance of many friends and professional brethren. While they do not complain of your having consulted what you consider as being your own interest in yielding to the influence under which you have acted, they must be permitted to say, that this influence, as you and the musical public cannot but feel, is of an arbitrary and oppressive character, and such as no theatrical manager or musical director has ever, *till now*, thought of exercising. What advantage can be derived by anyone from such an interference, is a question into which the Directors will not enter. But in separating from so many valued members of the orchestra, they cannot refrain from reminding you of the relation in which the Philharmonic Society has so long stood with the musical profession of London, and of the circumstances under which this separation has now been forced upon them.

At first sight it may appear that the directors of the Philharmonic Society were unnecessarily harsh with players whose only offence was in remaining loyal to a conductor endeavouring to provide them with more regular employment. Moreover, since Costa was responsible for keeping these players in training, and the management of the Royal Italian Opera for paying them a wage sufficient to enable them to devote all their time to orchestral playing, the Opera was in effect subsidizing the Philharmonic Society's orchestra, for the fees received by players from a society that gave only eight concerts a year would not keep even a poor orchestra together. Against this must be weighed the traditional policy of the Philharmonic Society, which did not exist for mercenary reasons, but sought only to uphold the dignity of the orchestral player and his art. In this ambition they had persevered and had accordingly always been able to rely on the goodwill of orchestral players, even, in fact, to expect some sacrifice of financial interest from them. In return the Society offered the players the highest prestige obtainable in this country. It was no mean return; engagements at provincial festivals were to be had outside the

THE NATIONALIST PERIOD

London season, and to be *persona grata* with the directors of the Philharmonic Society was a recommendation for such engagements. Therein lay the thinly-veiled threat in Hogarth's letter.

The only sensible way of overcoming the difficulty that had arisen was by compromise: the dates of the Philharmonic concerts could have been altered, or the times, or an arrangement could have been made with the management of the Opera for the loan of certain players on certain nights. This latter course was not possible in view of Costa's character. The directors of the Philharmonic Society well knew that Costa would not compromise. Had they not availed themselves of this very man to bring discipline into their own orchestra, and had they not lost him because he would not compromise with them on the choice of programme pieces? They had no illusions about the strength of will of a dictatorial conductor, but Hogarth's action in threatening the men whose livelihood condemned them to suffer under this arbitrary and oppressive will looks mean and shabby, for it was taken not to further a belief in freedom so much as to express the displeasure of those whom the players' loyalty had thwarted. Nor is it fair to blame Costa entirely for lack of goodwill; Sir George Smart, in insisting that it would be fatal to change the Philharmonic dates from Mondays, was equally stubborn.

There was need of a change of mind in the Philharmonic Society, but not for a change of principles. The advancement of the art of music and not of private financial interests remained its object, and all around were rival musical ventures organized for profit advertising their superior claims to leadership in musical life. Jullien's Promenade Concerts and Dr. Wylde's New Philharmonic Society had been sufficient to point to others the way to financial success in concert-giving. The first essential was a concert hall large enough to hold an audience that would cover the cost of an orchestra with a little profit to boot. The Hanover Square Rooms, ninety-five feet by thirty-five feet, were designed for eighteenth-century conditions, when orchestras were small and patrons wished the audience to be

MANNS AND GROVE AT CRYSTAL PALACE

exclusive; but the middle of the nineteenth century saw orchestras growing, and the Hanover Square Rooms were becoming insufficient, not only on account of the impossibility of containing an audience large enough to defray the cost, but also on account of their unsuitability for the larger volume of tone composers were demanding. A new and larger hall would offer a good return to investors, and so a limited liability company with capital resources of £40,000 built St. James's Hall, one hundred and thirty-nine feet long by sixty feet wide, in 1858.

Difficulties were encountered in the erection of the building, in consequence of which the cost rose to £70,000, and this had to be recovered by charges for the use of the building. It was therefore essential that the hall should be fully occupied every day. Monday was for some reason an unpopular day for concerts and few bookings were made for that day. In consequence, the owners of the building, led by their most important shareholders, Messrs. Chappell & Co., decided to organize a series of concerts to occupy the hall on Mondays. After a preliminary failure under Benedict, the management reorganized the venture under the happy title of 'Monday Popular Concerts', and on the advice of J. W. Davison restricted the programmes to classical music. The type of programme was similar at first to that of John Ella's Musical Union; and Davison wrote annotated programmes for the 'Monday Pops' similar to those he had written for the Musical Union. The coming of the analytical programme was, in fact, of much importance at this time,[1] for the attraction of the popular concert was partly in the satisfaction of an educational thirst much in evidence from 1851 onwards. Indeed, a good working impression of the trend of public taste for music can be had from a comparison of Davison's notes for the Monday Popular Concerts and Grove's notes for the Saturday Evening Concerts at the Crystal Palace. The 'Pops' were chamber music and the

[1] To Edinburgh belongs the credit for having first introduced annotated programmes.

Palace concerts orchestral, but they had the same educational trend and the same remarkable popularity.

Like the 'Monday Pops' the Crystal Palace concerts were organized by the proprietors of a building. When the Great Exhibition of 1851 came to an end, a company was formed to purchase the Crystal Palace, remove it from Hyde Park to a site on Sydenham Hill, and exploit it as a great centre of entertainment. A young architect named George Grove, who had had much experience in metal railway bridges and lighthouses, had become secretary of the Society of Arts in 1850, and in that capacity had had much to do with the erection of the Crystal Palace. In 1852 Grove became secretary of the new Crystal Palace Company, and by that chance came to have a most important influence on British music, for Grove was an amateur musician with a breadth of outlook unequalled among his professional musical friends; he knew how to employ the knowledge of others; he was able to collect a wealth of expert knowledge into his famous *Dictionary of Music and Musicians*, organize a first-class educational establishment in the Royal College of Music, and popularize good music at the Crystal Palace to such an extent that by 1880 it was to that unfashionable locality that music-lovers flocked to hear the best orchestra in London, and where British composers looked for a chance to introduce their works to the public.

It was done with remarkable speed and business efficiency. The first band employed at the Crystal Palace was conducted by a German bandmaster named Herr Schallehn and comprised sixty-two brass instruments, a piccolo and two E flat clarinets. To this band came a young German military bandmaster named August Manns as sub-conductor. He was unhappy under Schallehn, left the band after a few months, but returned the following season as conductor. The quality most in his favour with Grove is shown in the letter quoted on page 205.

Almost immediately after taking charge of the Crystal Palace band, August Manns conducted a Saturday concert in the Bohemian Glass Court of the Palace, and soon the type of band employed at these concerts was changed to an orchestra of 16 first violins, 14 second violins, 11 violas, 10 'cellos, and 10 double basses, with single wind. The Saturday programmes generally consisted of two overtures, a symphony, a concerto or some smaller orchestral pieces, and four songs. The length of this programme was only about half that of a Philharmonic concert, but this was an advantage, for these latter, as Wagner rightly said, were much too long and could not be adequately rehearsed in the time available. Manns changed all that: his wind players and some of his strings were regularly employed in the Crystal Palace band, and were in constant rehearsal under his baton. This was a factor of supreme importance, which the Philharmonic Society of London had never fully appreciated, and the progressive policy carried out by Grove and Manns at the Crystal Palace owed its success primarily to Manns' daily contact with the same players.

Hallé, as we have seen, fought for the same cause in Manchester and was able to establish there a permanent orchestra of eighty players, but at the risk of overworking himself and his men to such an extent that artistic results sometimes suffered. It remained for the Crystal Palace Company to exploit successfully the demand for good music in London without allowing the standard either of programmes or performances to decline.

It was done by a careful study of the public and the use of every means to make good music attractive to them. Saturday was the day when the largest crowds were to be expected, for it was the workers' half-holiday. Not all the workers spending Saturday evening at the Crystal Palace would be likely to go there for the music alone, but they would sample it along with the other entertainments just as the visitors to the Great Exhibition had done. The workers, too, were people of a wide variety of types, for besides heavy manual workers, the innumerable clerical workers with whom London abounded were all equally free on Saturday evenings, and, although the taverns still drew a good proportion of the people, there were great numbers of wage-earners whose religious principles led them

THE NATIONALIST PERIOD

to avoid such evils, and to seek sources of entertainment for themselves and their families where liquor was not in evidence. These people had perforce to avoid the theatres, and chose musical performances as the best type of entertainment available in the social environment they knew.

It did not rest here. Many who attended the Crystal Palace Saturday Evening Concerts were people of superior social standing whose genuine interest in orchestral music took them to Sydenham because the West End concerts did not give them all they wanted. In addition to Schubert and Schumann, Liszt, Brahms, Wagner, and Smetana works could be heard there. The public, however, can be forgetful, and Manns was particularly sensitive to any implication of neglect. In 1893, when Hans Richter conducted a performance of Smetana's *Vltava*, it was claimed to be new to this country. Manns wrote to the Press:

> The musical critics seem to have been under the impression that the above-named Symphonic Poem had never been heard in England before last Monday, when it was included in the programme of the Richter Concert. Kindly examine the two programmes of our Crystal Palace Saturday Concerts of 1881 and 1882, and you will find that I introduced *Ultava* and *Vigsebrad* as long as twelve years ago. I was much discouraged by the musical criticisms that followed my efforts (and my musical self) to bring these works of, at that time, a perfectly unknown Bohemian composer before my Crystal Palace audience, and I must add here that I did not fare much better in Glasgow, where, in 1882, I gave *Die Moldau (Ultava)*.

Such things disturbed Manns more than they need have done. He was hyper-sensitive — those who saw him conduct say that he was 'a bundle of nerves' — and Grove had constantly to act as his protector from the shafts of criticism; even at times try to allay Manns' unjust suspicion of his friends. In a letter to the critic Joseph Bennett in 1876, Grove says:

> I want to ask a kindness of you. Manns is in a terrible state of grief owing to various remarks in the papers

MANNS AND GROVE AT CRYSTAL PALACE

> recently which seem to give me more credit than is due — or rather to give him less — in reference to the Saturday Concerts. He urges that I am spoken of as if the choice of programmes, and the excellence of the execution, and the entire success of the concerts were due to me. I can't see the inference, but he does, and is terribly hurt and distressed. He is over-sensitive, but, on the other hand, he is so able and devoted, and has done so much more for music in England, that I should be very glad if he could be relieved in some way. He urges me to write to the papers, but this I am determined not to do. But it occurs to me that you could easily say something in your next notice that would heal the wound, and I am sure you will be glad to do so, both for my sake and his. I have written in the same sense to Ryan and J.W.D.

But Manns purred loudly enough when he was stroked the right way, for he wrote to the same critic:

> I cannot refrain from telling you that the very kind comments on my conductor-doings in England contained in the *Daily Telegraph* of yesterday have given me new blood for new exertions on behalf of new productions of good music. I have been perfectly hungering for lines of this kind from you, and therefore cannot refrain from telling you, in my own undisguised way, that you have made me happy, and that I thank you heartily and sincerely for the valuable support you have thus given me in my art-pursuits.

Despite this love of praise and suspicion of criticism, however, Manns was a conductor far in advance of any other in his time until the coming of Hans Richter. Manns was a German who did not despise British music, and indeed worked exceptionally hard to bring the works of unknown British composers before the public. By 1880 the time had gone when a British composer was content to appear before the public with the modest cough of a minor European: he was becoming assertive. At what time this worm began to turn it would be difficult to discover now, but there is evidence of his wriggling in 1842, when several people complained to the directors of the Philharmonic Society that they had been disturbed by someone hissing the pianist Thalberg, who was playing fantasias of his own composition on themes from *Don Giovanni* and *La Sonnambula*, and that the offender (whose name they said was M-cf-rr-n) flirted loudly with his female friends during the exposition of these masterpieces. Not until ten years later did a work of Macfarren appear on a Philharmonic programme, but between 1855 and 1888 most of his orchestral works were heard. August Manns, too, performed many of Macfarren's works at the Crystal Palace.

It needed a British composer with sufficient popularity to be respected before an effective protest could be made against the preponderance of German taste. Arthur Sullivan was trained partly in Leipzig and partly in the Chapel Royal, but by a lucky choice of librettist he was able to establish himself in public favour as firmly as any foreign musician. Long before this, however, he had had his first chance at the Crystal Palace, where his overture to *The Tempest* was first performed by Manns in 1862, repeated, and taken up by Hallé in Manchester soon afterwards. Sullivan took up the cause of the British musician strongly. Sullivan had little opinion of Sterndale Bennett, despite his acknowledged scholarship, because of his conservatism. 'He was', said Sullivan, 'bitterly prejudiced against the new school, as he called it. He would not have a note of Schumann; and as for Wagner, he was outside the pale of criticism.' But Bennett died in 1875 and the Philharmonic Society had to choose a successor. *The Times* stated that the directors would appoint only a British conductor, and the situation became at once amusing because by that ruling Hallé, Manns, Arditi, and Benedict, all suitable men, were eliminated. Alfred Mellon, conductor of the Musical Society of London, was regarded as a likely candidate, but the choice of the directors proved to be Sir W. G. Cusins, a capable musician, but one who was suspected of having got the appointment through family influence. (His uncle, G. F. Anderson, had been conductor of the Queen's private band, and until his death in 1876 was a powerful influence in Philharmonic circles.) So the Philharmonic Society began to inspire ridicule among the superior, and to this was added a multitude of other petty trials. Soft passages in the music were interrupted by extraneous noises in St. James's Hall known to emanate from the Christy Minstrels in the Lesser Hall, and a smell of cooking from the kitchens beneath added to the distraction. As new orchestral societies were inaugurated, clashes of dates occurred more frequently. The British Orchestral Society ran counter to some of their plans, the Alexandra Palace Concerts interfered, and above all, the Crystal Palace Saturday Evening Concerts clashed with Philharmonic rehearsals. Eighteen players were common to both orchestras! Then the Carl Rosa Opera Company was formed and further inroads were made on the players' time. Deputies were now common at Philharmonic concerts, subscribers fell off, prices were raised, protests were received, and in the midst of this a voice spoke up to say that it disliked both the band and the new German school of music (Brahms and Wagner were now to be heard at the Philharmonic). Clearly the old Society had fallen on hard times.

Meanwhile rivals were growing stronger. The Great Exhibition and the Crystal Palace were not the only ideas that the Prince Consort had set going through the medium of the Society of Arts. There was in addition the National Training School for Music at Kensington, of which Sullivan was appointed Principal in 1876. In 1882 this institution was merged into the Royal College of Music with a wider and more permanent basis. Sir George Grove was appointed director. Grove, then, as secretary of the Crystal Palace Concerts, principal of the R.C.M., editor of a renowned *Dictionary of Music and Musicians*, and a frequent contributor to the Press, exercised an enormous influence for the good of British music. Composition still clung to German models, but the British musician had grown strong enough to assert his right to be heard. Provincial choral festivals produced oratorios and cantatas by British

THE NATIONALIST PERIOD

composers; few of them, it is true, of any great merit, but with such a demand existing for their works composers had opportunities for advancement in their art. At the Crystal Palace Manns produced new orchestral works by men not heard of before — Parry, Stanford, Mackenzie, Cowen and, later in the century, Bantock and Holbrooke. These were the men who were to carry the torch of British music into the twentieth century, and finally to enable British music to break loose from foreign trammels, but the way was hard. In 1901 August Manns retired, honoured but outclassed by younger men. He was knighted in 1903.

HANS RICHTER

It was not thought abroad that the disinclination of some capable British musicians to follow the latest trends in Europe might be due to a desire to get away from them. Why should they? Stanford's *Irish Symphony* did not come until 1887, nor Parry's *English Symphony* until 1889, and when they did they were by no means free from German influence. The truth is that the British were beginning to feel the need for a distinctive national style, but did not yet know how to cultivate it. They awaited, though they did not admit it, the arrival of a first-rate musical genius — one who could stand with Brahms and Wagner in the opinion of the world.

Meanwhile, the German nation was increasing tremendously in power, and in pride promised to outstrip even the British. The Franco-Prussian War of 1870-71 had proved the Germans masters of the French, and the fusion of their many independent states into one Reich had made them into a formidable economic unit which showed every sign of growing rich on a rapidly-contrived Industrial Revolution that might reasonably be expected to avoid the worst of the social horrors that had struck Great Britain in the previous hundred years. A new ambition was spreading in Bismarck's Reich, and in that ambition German artistic supremacy took on an enhanced significance. Naturally, any sign of supremacy over their great economic rival, Britain, was a source of patriotic inflation for the Germans, and the phrase *das Land ohne Musik*, used to describe England, became effective nationalist propaganda before the end of the century.

Britain, meanwhile, was divided into a small section of professional musicians who in their own interests resented German 'vanity', and a very large and influential section, from the Queen downwards, who wanted only the best music and knew this to be German. Of these, two are especially worthy of

THE NATIONALIST PERIOD

mention: Walter Bache, who worked hard for the cause of Liszt's music, and Edward G. Dannreuther, who supported the cause of Wagner. Dannreuther conducted no less than nine orchestral concerts in 1873 and 1874 for the London Wagner Society. The object of this society (besides giving performances of Wagner's works) was to raise funds to assist towards the costs of the Bayreuth theatre. Some indication of the spreading fame of Wagner may be seen in the way such societies sprang spontaneously into being all over the world, for in addition to those started in Germany there appeared Wagner Societies in St. Petersburg, Warsaw, New York, Brussels, Amsterdam, Paris, Cairo, Stockholm, and Milan. By this time Wagner had become so great a figure in music that, though critical opinion split on the rock of his genius, the mud-slinging in which many London critics had indulged in 1855 would, if repeated, have brought only disrepute to the slingers. Facts about his private life there were which would have caused a scandal had they been published, but these were not made public. Instead, controversy ranged round his music. One had either to be a Wagnerian or an anti-Wagnerian, and to be a Wagnerian meant that one accepted the premise that Wagner incorporated and outclassed all who had gone before him in the departments of music-drama and orchestration, and that the future of music lay along the lines he had laid. It was possible in those days for a reasonable and well-informed man to take such a stand, for Wagner had exhaustively shown how he had developed the form of the music-drama from the operatic achievements of his predecessors, and from the point at which Beethoven had arrived with the *Choral Symphony*. Added to this there was a recognizable continuity of growth in the technique of symphonic programme-music from the *Fantastic Symphony* of Berlioz, through the method of theme-transformation used in the symphonic poems of Liszt, to the method of the *leit-motif* employed by Wagner, whose style of development could be accepted as the final stage in the constructional problem of programme-music.

HANS RICHTER

Those who could not agree with Wagner and his new music were mainly classicists, upholding their belief in the continuance of the classical symphonic tradition as exemplified in Brahms. They saw a greater future for abstract music than for programme-music. Among such men were Parry and Stanford, whose influence was great not only as composers but as teachers. Even Brahms, however, was content in his symphonies with an early nineteenth-century orchestra. (If we except the use of a bass tuba in Brahms' *Third Symphony* and a third drum in his *Fourth Symphony* we find Brahms content with an orchestra similar to Beethoven's.) Indeed, in all his symphonies Brahms restricts his horns and trumpets to the notes of their natural scales (including the notoriously bad ones) although valve instruments were available. One can see in this a kind of enjoyable self-discipline similar to that employed by G. B. Shaw when he kept the Greek rules of dramatic unities in his play *Getting Married*. Mendelssohn and Schumann were content to restrict the scoring of their symphonies to the modern equivalents of Beethoven's orchestra, avoiding such lush-toned upstarts as the harp, the *cor anglais* and the bass clarinet. Even those music-lovers who felt that the value of abstract music must ultimately be greater than that of programme music knew that the work of Wagner could not be ignored, because in it was to be found the highest development of orchestral sense known at that time.

Wagner was busy with the establishment of his Bayreuth theatre during the 'seventies, and in great need of funds. When, therefore, the violinist Wilhelmj suggested a great festival of his works in London, Wagner agreed, for London was still regarded as a sort of Klondyke by continental musicians. Wilhelmj was emphatic that Wagner's presence was essential for the success of the venture, but Wagner's life had been given up to Herculean struggles, and he felt that the strain of a London festival of the magnitude Wilhelmj suggested would overtax his strength. He therefore wrote to Hans Richter — who had gone to Vienna to produce *Die Walküre*, after conducting the whole

THE NATIONALIST PERIOD

of the rehearsals and performances at the opening Bayreuth festival of 1886 — 'For the accomplishment of my London scheme you are indispensably necessary to me; yes, without your help I really could not think of undertaking these concerts.' Richter responded with the loyalty he always showed to Wagner, and the London Wagner Festival was fixed to take place in the Albert Hall in May 1877, with Wagner and Richter as conductors. Edward Dannreuther helped to organize the British players needed for the huge orchestra, and directed some of the rehearsals.

This orchestra was the largest seen in London since Jullien's day, numbering one hundred and seventy players, led by Wilhelmj, and made up of 105 strings, 28 wood-wind, 29 brass and percussion, and 8 harps. The majority of these players belonged to this country, but some came with Wagner from Bayreuth,[1] as did the principal vocalists. The programmes included excerpts from *Rienzi*, *The Flying Dutchman*, *Tannhäuser*, *Lohengrin*, *The Mastersingers*, *Tristan*, etc. These pieces were confined to the first half of each evening and were conducted by Wagner. The second half of each concert was given up to excerpts from *The Ring of the Nibelungs*, conducted by Richter. It was obvious to the audience that Wagner was physically weaker than when he had last been in England — the driving force of the whole festival was Richter. Richter's conducting revealed a magnitude in Wagner's works which not even the composer could command. Londoners had seen conscientious conducting from Costa, brilliant conducting from Berlioz, but in Richter they saw greatness. First-class orchestral playing in England dates from 1877. The constitution of an orchestra, too, was settled; after 1877 Sir Charles Hallé's ophicleide continued to be heard in Manchester, it is true, but it was an obvious anachronism, and as for Sir Michael Costa's serpents! The last weak sections of the orchestra were made strong with Wagner's family of tubas.

---

[1] There are seventy foreign names printed on the programmes of the London Wagner Festival of 1877, viz. 57 strings, 10 wood-wind, and 3 brass, but of these many were resident in London.

HANS RICHTER

Unfortunate comparisons were made. It may have been true that Sir George Smart's directions to his men consisted mainly of repeating the phrase 'Pianissimo, gentlemen, pianissimo'. Orchestral players in those days were a tough lot, and Smart probably acted on the principle that if he looked after the pianissimos the fortissimos would look after themselves. Less plausible, however, is the story of the second horn player who deliberately used the wrong crook at a rehearsal. At the end the conductor — an Englishman, of course — said, 'Pretty well, pretty well', and turning to the second violins blandly added, 'There was something wrong with one of you, but it will be all right at the concert'. Most of these stories came from orchestral players (always the severest critics of conductors), but their stories of Richter are very different. The most famous is that of the horn player who declared his part to be impossible, whereupon Richter borrowed his instrument and played the passage correctly.

In consequence, Richter's most stalwart champions were among those who played under him. 'My colleagues', he called them, and right up to the time of his farewell speech to the London Symphony Orchestra in 1911 he deprecated any suggestion that conductors are heaven-born. He presented himself to his men as a master craftsman — one with a complete knowledge of the orchestra — and his men knew this claim to be true. He got better results from them than any other conductor had previously done, and was completely honest. On one occasion he was conducting Brahms' *Academic Festival Overture*, as usual from memory, when he made a slip and beat the *alla breve* passage two bars too soon. At once Richter stopped the orchestra and addressed the audience; he explained simply that the error was entirely his fault, and restarted the overture from the beginning.

Such was Richter's character, and he was destined to spend over thirty years as a conductor in this country. The impression made at the Wagner Festival of 1877 had as a result an invitation to conduct further concerts in London. According to

THE NATIONALIST PERIOD

Richter himself, 'Hermann Franke, an excellent artist and an estimable man, was really the founder of the Richter Concerts. Franke suggested that I should come to England from time to time. I was delighted with the English musicians, and I told him that I should like to conduct concerts in London'.

The first Richter Concert took place in St. James's Hall on May 5th, 1879, with a well-mixed programme of classical and romantic masterpieces. Richter's conscientious readings of the scores — if reading is the right word, for he knew them so well that many (especially those of Beethoven) he conducted from memory — enhanced considerably the already great reputation his conducting of Wagner's scores had earned. Musical criticism had veered right round between 1855 and 1877 in Wagner's favour, but it had not improved in quality. London critics there were who still found it impossible to praise one party without odious comparisons. Richter was praised at the expense of British musicians.

'Herr Richter has not under him here, as at Bayreuth, the picked artists of an Empire, nor has he had almost unlimited rehearsals, but he is one of the men who, in a certain sense, falsify the dictum that you cannot grow grapes on thorns. If Napoleon's presence with his troops was worth, as said Wellington, an army corps of 20,000 men, what is the value of an orchestra with this emperor of conductors? We cannot appraise it, but we can feel the influence of Richter's supreme mastery; of his all-embracing *coup d'œil*, of his perfect resource, and, not less, the confidence with which he must inspire his followers. Here Richter is a "conductor" of a verity, and we are glad to have him amongst us as an example. Many of our own conductors have been sitting at his feet this week, and we trust that the fidgety among them will fidget less; the convulsive become calmer; the uncertain more assured; the feeble stronger; and, we had almost said, the led ones themselves take the lead, though that would, perhaps, be a change for the worse. As respects Herr Richter's reading of Wagner's music, nothing need be said after the experience of the "Wagner Festival", and

HANS RICHTER

with reference to his treatment of Beethoven, we are chiefly to appraise the discretion which avoids all false interpretation. He brings out into fuller relief that which is obvious in the score, but he does not treat the great master's music as an obscure text given him for emendation. This is one of the great Viennese conductor's greatest recommendations to us conservative English.'

So from the beginning Richter became a goad to nationalist sentiment in this country without having any wish to be so. Much as a large good-natured dog will sometimes inspire a noisy yapping from impotent but vociferous poodles, so Richter stood in the midst of the British nationalists — dignified, even genial. The Richter Concerts filled St. James's Hall year after year, and in 1882 and 1884 he conducted German operas in London, introducing *The Mastersingers* and *Tristan* in their complete versions. In the latter year he accepted the post of conductor of the Birmingham Festival for the following year, 1885, and one of the poodles yapped: 'If it is true that Richter has been, or is to be, offered the Birmingham Festival, I think it is an affront to all of us English. . . . I should certainly have considered it an honour if they had offered me the festival, whether I could have undertaken it or not. But it is not entirely selfish, for not a thought of envy or regret should I have felt if Cowen, Stanford, Barnby, or Randegger (who is one of us to all practical purposes) had been selected. They would have done the work well.'

Perhaps it is unwise to describe Sir Arthur Sullivan as an impotent and vociferous poodle, but he was barking from a very undistinguished kennel. Richter was not perturbed; he probably thought more of the opinions of those who had direct experience of his work, such as that given by his trumpet player, Walter Morrow,[1] in an interview with a representative of the *Muscial Times* in 1899:

> Yes, I am proud to say that I have been a member of Dr. Richter's orchestra since 1879, the date of his first

[1] Walter Morrow is important as the inventor of the British version of the Bach trumpet.

THE NATIONALIST PERIOD

concert in London. During this long period of twenty years I have seen many changes in the personnel of his orchestra. A number of his earlier players have departed this life, among them Svendson, Trout, Ellis, Hurley, Geard, and Catchpole. All these were highly esteemed by Dr. Richter — indeed, he esteems and is esteemed by every member of the band; or he makes them believe he does, and this is a very great matter in the management of an orchestra.

I perfectly remember the first impression he made on us orchestral players at his first rehearsal in St. James's Hall; his conducting was a revelation. His many high qualities stamped him at once in our minds as a master conductor: an impression that has not changed, but greatly strengthened. Some prophesied that he, being new to us, only showed his better side, and that time would unveil the objectionable side. This prophecy has not been fulfilled; he is the same Hans Richter, possessing the same grand qualities as of old.

I have sometimes been asked to describe those qualities. It is difficult to enumerate them; they can be felt, but cannot be adequately described. He has an imposing presence; a generous, genial manner; a wonderful self-command; a prodigious memory; a profound knowledge of scores; a practical knowledge of instruments, particularly the trumpet and the horn — the technical difficulties of these two tender instruments are rarely understood. The players know how kindly and sympathetically Dr. Richter nurses them; at the same time he gets everything possible out of them.

In conducting an orchestra Dr. Richter naturally beats time with a baton, and his beat is unmistakable; but his power is not there — it is in his eye and in his left hand. What a wonderfully expressive left hand it is! And he seems to have every individual member of the band in his eye; he misses nothing, and we do not seem able to escape from the influence of that eye for a single moment. No mistakes occur without the influence being severely felt. At one concert I was under the impression that I had

accomplished the feat of making a mistake which he did not detect. We were playing *The Flying Dutchman*: I let my thoughts wander for a moment and at one passage entered just one bar too soon. I only touched one note and stopped; the Doctor's eyes were on the second violins. I said to my colleague, 'He has not caught me this time'. The Doctor took no notice of me, and we finished the first part of the concert. During the interval my colleague and I were chatting together quite cheerfully, when I felt a touch of a hand on my shoulder, and heard the question, 'What happened in the Dutchman, Mr. Morrow?' I magnanimously acknowledged that 'It was a little carelessness on my part, Doctor'. 'A little what?' he said. I repeated, 'It was just carelessness'. That was a word he did not know, at least in English; he evidently thought it was some miserable subterfuge, for he walked away shaking his head and saying: 'No, no. It was a wrong note.'
I know no conductor who can get so much out of orchestral players. At rehearsals he gets through a large amount of work, but there is never any friction, never any irritability or irritation. He stops a great number of times, but always for a satisfactory and well-explained reason, and those explanations are often very amusing. The violoncelli, for instance, have a passage in *Tristan* which should be rendered with much warmth of expression. The rendering, however, though generally correct, is cold. This elicits the remark: 'Bravo, celli! Quite correct, but you play like married people; a little more like the young lovers, please.' At the end of a three hours' rehearsal we disperse with the feeling that a great deal has been taken out of us, but that the operation has been pleasant and agreeable. At the close of a concert we realize that the strain has been very great; but the artistic renderings under Dr. Richter's baton, and the kind appreciative words he never fails to address to his orchestral players always arouse feelings of admiration and esteem for the greatest living conductor.

Some present-day orchestral players, like Bernard Shore and Thomas Russell, can express themselves more lucidly than Walter

THE NATIONALIST PERIOD

Morrow, but it is to be doubted if they have ever felt greater admiration for a conductor than Morrow did for Richter.

Richter spoke little at rehearsals; that personality which drew from his men the utmost of which they were capable was as characteristic of his rehearsals as of his concerts. A distinctive feature of Richter's at rehearsal was the start or restart of a tricky passage, when Richter would stand impassive, stick held aloft, in a long pause which had the effect of compelling concentration on the phrase that was to come. The greatness of Richter, men felt, was not in himself but in the music he was conducting. He was a channel through which that greatness had to pass, just as his players were. Conversely, Richter had no use for music that lacked magnitude. It is not to be inferred from this that he was stolid or pompous: while it is true, as one of his admirers said, that 'no other could unfold that splendid banner of tone with which *The Mastersingers* overture opens', it is equally true that he could always 'let it disperse later into its myriad smiles without losing anything of its greatness, and bring its overwhelming climax, without strain or loss of ease, back to its great simplicity'.

Richter's greatest work in this country was done in Manchester as conductor of the Hallé Orchestra. Sir Charles Hallé died in 1895, and after a disappointing period under Cowen Richter was appointed in 1899. There he remained until 1911, when failing sight compelled him to retire. There he enjoyed the full support of the Hallé Society, and the advantage of living in a city in which the foundations of a high standard of musical education had been well and truly laid, for Sir Charles Hallé had, besides the Society that bore his name, succeeded in founding the Royal Manchester College of Music to fill his auditorium with discerning listeners and train players who in due course could fill the ranks of his orchestra. Richter's work was backed by his leader, Dr. Adolph Brodsky, chairman of the Hallé Society, professor at the R.M.C.M. and leader of a string quartet second in fame only to that of Joachim.[1] Like

[1] Carl Fuchs, Richter's leading 'cellist, also had a quartet worthy of notice.

HANS RICHTER

the Philharmonic Society of London, the Hallé Society existed for artistic, not financial aims, and this spirit was further disseminated by Dr. Brodsky, who devoted the proceeds of his chamber-music concerts to the needs of students at the College. It is doubtful if any other city in England had so generous a disposition towards music as Manchester in the Edwardian period. There Richter was able to devote himself wholeheartedly to his craft[1] even as he had done abroad, and the Hallé Orchestra showed corresponding results. At last we had an orchestra in England that would bear favourable comparison with those abroad — the Hallé wood-wind was especially fine. Backing it was a musical community loyal to a fault. The Henry Watson Music Library of 16,000 volumes,[2] presented to the city by a Salford benefactor, made for a better-informed public than was possible in many other cities, and a succession of critics on the staff of the *Manchester Guardian* — Ernest Newman, Arthur Johnstone, Samuel Langford, and Neville Cardus — has in the present century done much to revolutionize musical criticism. The Hallé Orchestra continued the policy of its founder in supplying the demand for orchestral players in some thirty or more towns in the north, and in the summer holidays many of the same players could be seen leading the pier orchestras in popular resorts in North Wales and along the Lancashire coast. The Hallé Orchestra was a people's movement.

Richter's baton, however, was no magic wand, to enliven a dull work with a puckish spirit. Under Richter a Schumann symphony would be noticeably slighter than one of Brahms. From Richter's point of view, Sir Thomas Beecham's spicing up of Schumann would be a form of artistic alteration. Naturally, in a city so well informed as Manchester, Richter had his critics. He was solely responsible for the choice of music in the Hallé programmes, and only Sir Henry Wood has succeeded in

---

[1] Except opera, which Richter could not persuade Manchester to put on a permanent basis.
[2] It now holds (1945) over 40,000 volumes.

THE NATIONALIST PERIOD

pleasing everybody in orchestral circles. Richter thought in terms of musical taste — not communal taste in music. His readings of Wagner were authoritative, and he did valiant work on behalf of Richard Strauss; he had not the same warm enthusiasm, however, for the music of Sibelius, and although, in accordance with the Hallé tradition, he gave electrifying performances of Berlioz, he had little regard for other French composers, and none at all for César Franck. A performance of Debussy's *Prelude à l'après-midi d'un faune* was carefully rehearsed and conscientiously performed, but failed to satisfy lovers of Debussy's style. Richter's approach to all music was unconsciously Germanic, and his severest critics in Manchester were those who wanted a wider scope of music. Sir Thomas Beecham introduced the works of Delius to Manchester, and a short-lived club called the French Concerts, introduced Saint-Saëns; Sir Henry J. Wood managed after a hard fight to establish the Brand Lane Concerts in popular favour, but Manchester remained nevertheless faithful to the Hallé Society. The difference between Richter and Wood was the difference between London and Manchester. In *My Life of Music* Sir Henry states that he was never greatly impressed by the Richter Concerts at St. James's Hall, but was thrilled by those he heard under Richter in Vienna, Berlin, Munich, and Bayreuth. Sir Henry suspects that this was due to German contempt for British musicians shown at that time, but Manchester people found London orchestras less satisfactory than their own, although Richter maintained that the London men were the best readers, and certainly there is no evidence from Richter's British players that he treated them with contempt — quite the contrary. The explanation lies in the closer affinity of Germanic temperaments than of mixed Germanic and Anglo-Saxon temperaments. There was a strong stiffening of German merchants behind the Hallé Society.

Elgar's admiration for Richter is well known. He preferred Hans Richter to all other conductors. In his early years Elgar had had great difficulty in getting his music performed — his

HANS RICHTER

friends in the Severn Valley had taken to his choral music, but he had to look to London for performances of his orchestral works. In 1897 Elgar wrote to the secretary of the Philharmonic Society of London offering them an orchestral work for performance, but would not submit the score for approval; the Society therefore could not accept Elgar's offer. Richter performed the *Enigma Variations* in London in 1899 with every success, however, and from that time Elgar's loyalty to Richter dates. The successful first performances of *The Apostles*, *The Kingdom*, and the *Symphony in A flat* were as surely due to Richter's personality as was the failure of *The Dream of Gerontius*. The nobility of Elgar's themes and the sonority of his orchestration appealed to Richter, who saw in them a continuance of the magnitude of the romantic tradition. The retirement of 'Hans Richter, Mus. Doc., True Artist and True Friend' (as the dedication of the *Symphony in A flat* runs) was by nobody in this country more keenly felt than by Sir Edward Elgar.

MUNICIPAL MUSIC
UNDER SIR DAN GODFREY

A YOUNG man named Henry J. Wood was once sent by his father to Llandudno to hear the orchestra of Rivière, who was at that time conducting concerts in a hall on the Esplanade in that town. Here the young man got the shock of his life. As he took his seat he saw an elderly gentleman seated in a gilded arm-chair facing the audience. He was elegantly dressed in a velvet jacket, on the lapel of which reposed a spray of orchids more fitted, Wood says,[1] for a woman's corsage. He held a bejewelled ivory baton in his hand from which dangled a massive blue tassel. This he wound round his wrist. He bowed ceremoniously to the audience and tapped loudly on his golden music-stand. Still seated, he began the overture to *Mignon*. After two bars, a hoarse voice from the side of the orchestra said, 'Six beats in a bar, please!'

Wood escaped, and went to the pier concerts where Bartlet was conducting a British orchestra. Here was something completely different: a packed house listening to a good, if rather small, orchestra. The leader was Arthur Payne, who played the *Andante* and *Finale* from Mendelssohn's *Violin Concerto* with such beauty of tone that Wood made a mental note that if ever he should want a leader he would offer the post to Payne.

This experience illustrates a point of some importance in the development of orchestral policy taking place at that time in holiday resorts. The 'refined' small orchestra with its foreign conductor had for years been a feature of such holiday resorts as catered for middle-class visitors. M. Rivière was an extreme example of a type dear to the hearts of good Victorians, but the naughty 'nineties brought a reaction against such innocuous entertainment, and resorts that saw fit to provide a better standard of orchestral playing were growing in favour. Most

[1] Sir Henry J. Wood, *My Life of Music*.

MUSIC UNDER SIR DAN GODFREY

noteworthy of such towns is Bournemouth, which has played a worthy part in the advancement of British music, and has stood for half a century as an example of what can be done by a discerning corporation to enrich itself and the musical life of its inhabitants. Bournemouth has been fortunate in having had the services of an ideal organizer in Sir Dan Godfrey, but as Sir Dan never failed to admit, his success could not have been attained without the goodwill of a particularly far-seeing municipality.

To prevent a possible misunderstanding, it must be emphasized that Bournemouth residents have little sympathy with the political doctrine that advocates national or local government control in preference to that of private enterprise; but special circumstances obtained in Bournemouth in the late nineteenth century, and music offered a way out of the business men's difficulties. At that time Bournemouth profits were dependent on the attraction of retired people — mostly from the army, navy, and civil services — and wealthy invalids who sought the healing effect of the sea breezes and the odour of Bournemouth's extensive pinewoods. Bournemouth was as Thomas Hardy described it: 'A city of detatched mansions; a Mediterranean lounging-place on the English Channel. . . . The sea was near at hand but not intrusive, it murmured and Angel Clare thought it was the pines, the pines murmured in precisely the same way and he thought they were the sea.' Sports and other virile forms of entertainment had few advocates in Bournemouth. The place, in fact, was deadly dull, and a large glass pavilion housing various exhibitions after the manner of a miniature Crystal Palace failed to pay its way. The owners looked for a buyer of white elephants — it is at such times that property-owners are in favour of municipalization.

The Bournemouth Corporation did not act blindly in the matter, however, for, although music seemed to offer a way to relieve the dullness of their town, it would have to be improved considerably if the upkeep of a pavilion was to depend on its popularity. There had since 1876 been an Italian Band of

THE NATIONALIST PERIOD

sixteen players in Bournemouth, all of whom had been in the Italian army and wore the uniform. They were paid, according to Sir Dan Godfrey,[1] 'by public subscription', which is probably the Bournemouth way of saying that they passed round the hat. In 1892 Signor Bertini was engaged to provide the first Corporation Band of twenty-one players, most of whom had belonged to the Italian Band. In 1893 the Corporation took over the lease of the Winter Gardens (the glass pavilion already mentioned) and appointed Daniel E. Godfrey municipal bandmaster with a wind band of thirty players. His duties were to provide music daily on the pier, but a number of his players were 'double-handed', i.e. they could play two instruments. Thus Godfrey was able to provide a small orchestra for use in the Winter Gardens pavilion in addition to military band concerts on the pier.

Dan Godfrey, who came of a long line of army bandmasters, was known to some Bournemouth residents, but special qualities he was to develop in their town they could not have known at that time. From the first his orchestra and band were popular. The first programme may be quoted to show the kind of music on which he based this popularity:[2]

March	The Standard Bearer	*Fahrbach*
Overture	Raymond	*Thomas*
Valse	Je t'aime	*Waldteufel*
Ballet Music	Rosamonde	*Schubert*
Russian Mazurka	La Czarine	*Ganne*
Entr'acte	La Colombe	*Gounod*
Selection	The Gondoliers	*Sullivan*

The success of Godfrey's band was such that five thousand people attended this concert, and public support throughout the summer continued to be such that the Corporation decided to retain Godfrey and a double-handed nucleus of twenty-four players throughout the winter season. Next year, 1894, Godfrey was appointed resident Musical Director of the first permanent Municipal Orchestra in Britain.

[1] Sir Dan Godfrey, *Memories and Music*, 1924.
[2] Given on May 22nd, 1893.

MUSIC UNDER SIR DAN GODFREY

Having once established a reputation for good music, however, the Corporation resorted to the not uncommon business practice of reducing costs while still hoping that the reputation already gained would keep up the demand. For the winter of 1894 Godfrey's forces were reduced to a wind band of eighteen players. The trick failed to work. The following summer the original type of double-handed band was restored, but increased to thirty-three players, all of whom were retained during the following winter. Godfrey was able to strengthen considerably his string section, institute special classical concerts every Tuesday during the summer of this year, and pass on to symphony concerts during the winter. So well supported were the first forty symphony concerts that a further twenty had to be added, and these served to fill the whole winter season from October to May. In 1896 Dan Godfrey was appointed General Manager of the Winter Gardens while still retaining the post of Musical Director. March 1st, 1897, saw the one hundredth symphony concert given in the Winter Gardens (the symphony was Tschaikovski's *Pathetic*) and the first phase of Godfrey's epoch of municipal music may be said to have ended. In that year was held the first Bournemouth Musical Festival, August Manns sharing with Godfrey the honours of conducting.

Certain definite traits of taste had now revealed themselves in Bournemouth audiences, notably a fondness for Brahms. This is said to be due largely to the enthusiasm of Mr. John B. Camm, secretary of the 1897 festival, who had long been accustomed during previous residence in London to attend the Crystal Palace Concerts, and had there developed a passion for Brahms with which he successfully infected his friends in Bournemouth. Godfrey's concerts appealed to Camm even more than those of Manns, for it had long been his habit to winter on the Continent. Godfrey's orchestral concerts, however, kept him in Bournemouth throughout the year and even drew him into practical efforts for their improvement, for in addition to acting as secretary to the Festival Committee he gave Dan Godfrey the free use of his extensive library of orchestral scores,

THE NATIONALIST PERIOD

and at his death bequeathed them, through Godfrey, to the town. The Camm collection of scores at the Lansdowne Public Library in Bournemouth is unsurpassed among municipal free libraries in this country except for the Henry Watson Library at Manchester.

The magnitude of Bournemouth's achievement at this date can best be gauged by comparison with another experiment under an equally able conductor. Between 1897 and 1900 a promising situation appeared in the North of England and collapsed. New Brighton, within easy reach of Liverpool by the Mersey ferry, set out to provide the attractions of Blackpool in the Mersey area. With this object a company was formed to erect a tower after the Blackpool model, with adjacent ballroom and gardens. Granville Bantock was appointed musical director of the new venture, his duties being to provide music for dancing in the ballroom for five or six hours each day excepting Sundays.

Bantock, however, was a musician with ambitions beyond dance music. Already he had shown the way his genius was leading by a series of compositions that included a symphonic poem *Jaga Naut* later to be performed by the Philharmonic Society of London, a cantata *The Fire-worshippers* to words from Moore's *Lalla Rookh*, the overture of which Manns had produced at the Crystal Palace, and a one-act opera *Caedmar* which had had three or four performances at the Olympic Theatre by the ill-fated Signor Lago's Opera Company, Bantock conducting. He had also published many songs. Why, then, is Bantock to be found conducting trivialities at New Brighton? The fact is that facilities open to a young conductor on the continent did not exist in this country. In his biography of Sir Granville Bantock, H. Orsmond Anderton describes how he spent an evening with Sir Edward German, then a young man, walking up and down Hanover Square discussing whether German should accept or reject an offer to conduct a theatre band. In the end, German decided to take it, and that led to his friendship with Sir Henry Irving and composing the incidental music to *Henry VIII*, which established his reputation in the theatre and largely conditioned his future style of composition. Bantock, too, had been able to find no more suitable employment than conducting musical comedies on tour; like these, the New Brighton appointment offered no more than a means of keeping the wolf from the door.

Bantock made the best of the situation; he carried out his duties conscientiously and satisfied the management; he began to think, however, of ways by which he might improve the standard of music at New Brighton. There was not much that could be done at first, for the buildings were not fully erected in 1897; but he took the strain of his band, and tried to sum up the attitude of the management. A young organist, visiting New Brighton with his choir at an annual outing, entered a partly-erected portion of the Tower buildings and heard the sound of the *Tristan Prelude* coming from behind a tarpaulin. It afterwards transpired that Bantock was using the orchestra's rehearsal time to practise this music instead of the dances that they were to play that evening. So did the Bantock experiment start.

On the committee of management one man — by name de Ybarrondo — advocated a higher class of music than that being given. Although the Tower set out to attract the heartier types of holiday-makers, it was known from the experiences of Jullien and Manns that music of a high quality attracts a better class of audience than dance music — an audience not given to drunkenness and such other vices as bring a place of entertainment into disrepute. Bantock and de Ybarrondo were not puritans, but they were of the opinion that good music, besides being desirable in itself, would ultimately benefit New Brighton.

Gradually Bantock began to improve his Sunday programmes, and made suggestions for the improvement of his weekday dance band with the object of adapting it to a more varied choice of indoor music. Dancing was less popular in the afternoons than in the evenings; afternoon programmes could

THE NATIONALIST PERIOD

therefore be mixed. The type of programme that resulted can be seen from one quoted by Anderton that took place on Monday, May 5th, 1898:

Coronation March	Henry VIII	*German*
Waltz	Moonlight on the Rhine	*Vollstedt*
Polka	Chin Chin Chinaman	*Kiefert*
Waltz	Tres joli	*Waldteufel*
Selection	The Geisha	*Jones*
Waltz	Blue Danube	*Strauss*
Invitation à la Polka		*Thomé*
Galop	Troika Race	*Damare*

The next stage was special concerts for which the orchestra was augmented and a symphony included in the programme. The following took place on Friday, June 3rd, 1898:

Overture	Egmont	*Beethoven*
Elegy for Strings		*Tschaikovski*
Symphony in C, 'Jupiter'		*Mozart*
Siegfried Idyll		*Wagner*
Hungarian Rhapsody No. 2		*Liszt*

For following Fridays popular symphonies were announced which included Dvorák's *From the New World*, Tschaikovski's *Pathetic*, Rubinstein's *Ocean*, a Grand Wagner Concert, and a cycle of Beethoven symphonies. Bantock had reached the contemporary Bournemouth standard of taste, and had he been content to continue at this level he might ultimately have created a demand for a permanent symphony orchestra in New Brighton worthy to stand with the Hallé Orchestra of Manchester. But Bantock had no ambition to be a pale reflection of Richter: his aim was to make New Brighton famous for the music of living composers, including those of British origin, and to do so at a price that the working-class visitors to New Brighton could afford. By 1899 all his plans were laid. On May 28th of that year a Cowen concert took place, conducted by the composer (Sir Frederick Cowen was well known in Liverpool as conductor of the Liverpool Philharmonic Society). Other composers followed, all of whom conducted complete

MUSIC UNDER SIR DAN GODFREY

programmes of their own works. Stanford on June 25th, then Parry, Elgar, Corder, Wallace, German, Emile Mathieu (of Ghent), and Mackenzie. Bantock himself conducted a miscellaneous British concert, a Tschaikovski concert, and a Liszt concert. For other concerts he invited various well-known continental conductors: M. Camille Chevillard, of the Lamoureux Concerts, Paris, for a French programme; M. Sylvan Depuis, of Liége, for a Belgian concert; while a Liverpool conductor, Alfred E. Rodewald, conducted a Dvorák programme. Rodewald was a friend of Elgar (the first *Pomp and Circumstance* march is dedicated to him) and through Rodewald as well as Bantock the music of Elgar was made familiar to Merseyside audiences. Nearby Manchester specialized in the great classical and romantic composers, with Elgar and Richard Strauss accepted as followers of that line, while Liverpool was becoming famous as a centre for musical nationalism.

No less remarkable was the cost to the audience. Those who paid sixpence admission to the Tower grounds were admitted to the gallery free, or could reserve a seat for a further sixpence; for half a guinea one could buy a ticket for a series of eighteen Sunday afternoon concerts. The orchestra for these events was augmented to a hundred players, led by Vasco V. Ackeroyd, and the general ensemble was good. (The wind, brass, and percussion belonged to Bantock's permanent Tower band, and were therefore in daily rehearsal under him.) The committee of management, however, was lukewarm towards the venture, and their chairman openly hostile. Mr. de Ybarrondo — still Bantock's only champion on the board of directors — was ultimately driven to resign. Thus Bantock's three years' work was brought to an end, and the Tower continued its business career freed from artistic impediments.

At first sight it might appear that comparison between Bournemouth and New Brighton reveals only the difference in music under municipal and capitalist control, but there is more in it than this. The two towns catered for different audiences.

THE NATIONALIST PERIOD

Bournemouth's visitors were wealthy invalids, New Brighton's visitors hearty working-class people. These features governed the line of exploitation, for the demand for hearty entertainment, insignificant in Bournemouth, was predominant in New Brighton, and far greater profits can be made from sports, refreshments, and fun-of-the-fair devices than can be made out of orchestral concerts at half a guinea a season. New Brighton looked with admiration to Blackpool, where their vision of fat profits had never been dazzled by the glare of an artistic reputation.

Sir Dan Godfrey, by reason of long practical experience as manager and musical director in Bournemouth, could speak with authority on matters appertaining to municipal music. He understood not only the intricacies of organizing and conducting a first-class orchestra, but also local government opportunities, obligations to the law, the officials' duties, and the vagaries of council decisions and municipal elections. If a municipal orchestra is to function with any effective permanence all these things must be co-related. Sir Dan kept an eye on the musical activities of other towns besides experimenting in his own. How thoroughly he understood the snares and possibilities of municipal music can be seen in an address he gave to the members of the Incorporated Society of Musicians in 1923. Judging from his remarks on other towns on this occasion it is likely that Godfrey's opinion of the New Brighton experiment would be that Bantock tried to move too quickly. He gave that as the reason for Basil Cameron's failure to secure permanent results with his symphony orchestra at Torquay prior to 1914. Torquay had some features in common with Bournemouth, but Cameron tried to make the change from light music to symphony too quickly. It is a great mistake, in Sir Dan Godfrey's view, to underestimate the attraction of a cornet solo interposed at the point of greatest strain in a Sunday programme. Sir H. J. Wood in his earlier Promenade Concerts relied much on the popularity of cornet solos in those days, as did Sir Dan, and to this day there is in the Bournemouth

MUSIC UNDER SIR DAN GODFREY

Municipal Orchestra an excellent xylophone soloist, a successor to W. Byrne, one of Godfrey's original band, who, on his retirement in 1921, paid a compliment to his conductor that Godfrey treasured more than the after-dinner utterances of innumerable mayors. 'He's hot stuff', said Byrne, 'but he's just.' It is the sort of thing that Costa's and Richter's men have said.

Godfrey got results by hastening slowly. Low pitch, which came into use at Queen's Hall in 1895, Godfrey did not venture to use in Bournemouth until 1909 — the Corporation advanced money for the cost of new instruments, which the players repaid, less a discount of one-third, by weekly deductions from their wages. The next advance came in 1912, when the Corporation was prevailed upon to engage a separate military band, thereby relieving Godfrey's double-handed players for more orchestral rehearsals. His excellent results were obtained by much rehearsal of short programmes. He regarded one and a half hours' music as the limit that a seaside audience would stand on a pleasant afternoon. The programme formula he found most acceptable was an overture, a symphony, a concerto, and a novelty item. This for a full symphony concert. Those who preferred lighter fare could attend his Monday Popular Concerts (he dropped the name 'Classical' in favour of 'Popular'), where Godfrey's policy was to whet the appetite of the more serious-minded members of these audiences with single movements from typical symphonies, such as the finale of a Haydn symphony, the slow movement from Mozart's *Symphony in E flat*, the *Allegretto* from Beethoven's *Seventh Symphony*, one movement from Schubert's *Unfinished Symphony*, the *Scherzo* from Mendelssohn's *Scotch Symphony*, the *Largo* of Dvorák's *Symphony from the New World*, or the finale of Brahms' *Symphony in D*.

In addition there were attractive educational programmes offered to students and school children. Such things as *The Evolution of the Overture:* Gluck to Humperdinck; *The Evolution of the Ballet:* from Beethoven (*Prometheus*) to Borodine (*Prince Igor*); musical portraits of famous characters, e.g.

THE NATIONALIST PERIOD

Berlioz's Rob Roy, Litolff's Robespierre, and a programme of Shakespeare's characters. Special programmes for more advanced students included 'Dance', 'Symphony', 'Operatic Overture', and 'Concert Overture'. There were also national programmes to be had: British, French, German, Italian, Russian, and Scandinavian. All these were being regularly done in Bournemouth before 1924. In the matter of educational approach Godfrey learned much from Dr. Walter Carroll, who did great work as Musical Adviser to the City of Manchester.

Such progress was possible because Sir Dan Godfrey had made Bournemouth familiar with the orchestra as a part of their civic life, just as their museums and libraries are. It was a source of civic pride. A popular guide-book published in 1928 says: 'Bournemouth may justly pride itself on the fact that it was the first British municipality to establish a permanent band, and it may congratulate itself upon the result of its enterprise. Its Municipal Orchestra celebrated its coming of age in 1914, while in 1922 the honour of knighthood was conferred on its gifted conductor . . . Moreover, the Bournemouth authorities have found that the liberal provision of good music "pays". From time to time they have prudently increased their outlay upon it, so that the total expenditure on music is now about £20,000 a year. At the special concerts the foremost singers and instrumentalists appear, while the performance of works under the conductorship of their composers lends an additional interest to many of the concerts, and in the case of new works does much to forward British music in general.' Improvement went on after this date with the erection of a new pavilion having better acoustic properties than the old Winter Gardens Pavilion, until at the time of Sir Dan's retirement in 1935 a full symphony orchestra of seventy players was permanently available. At that time only the B.B.C. and the London Philharmonic Orchestras permanently employed an orchestra of these dimensions (the Hallé Orchestra players were engaged for a six months' season only). With such a record to his credit it is

MUSIC UNDER SIR DAN GODFREY

not remarkable that Sir Dan believed firmly that the future of British orchestral music depended on the extension of municipal support. Few would be prepared to dispute one statement at least that he made at a Congress of the British Music Society in 1922: 'It is perfectly obvious that private enterprise will never supply the needful in the matter of educational music.'

Few would be prepared to dispute it, but one did. Sir Thomas Beecham gave his views at the concluding banquet of this congress, and Sir Thomas is an uncompromising individualist, with a flair for after-dinner speaking as well as conducting. Sir Dan Godfrey felt it incumbent on him to reply, and did so in the columns of the *Daily Telegraph*:

> As to what has been done by municipalities, surely Sir Thomas cannot be ignorant of the large sums spent by the London County Council, and in other cities, to provide free public music for the masses; or the gratifying success of Mr. Appleby Matthews' orchestral concerts in Birmingham, which in the future will be backed by the City Council; or the efforts in Bradford, Leeds, etc. Does he consider the work carried on by the Municipality of Bournemouth for the last twenty-seven years of no value? Namely, that with the maintenance of a permanent orchestra of forty players (augmented when required) and the allowance of adequate time for rehearsal, sound performances have been given of between 700 and 800 important British works, at which more than a hundred of the composers have had the opportunity of conducting? The system has been to rehearse each new work four times — a point I wish to stress; at the second or third rehearsal the composer is invited to listen to his work, so that he may have an opportunity of judging its effects and note the faults, thus giving our young British writers an opportunity which has not been possible elsewhere (till the recent change in the application of the Royal College of Music Patron's Fund) of comparing their written ideas with actual effect. I naturally felt that Sir Thomas's sweeping remarks were most unjust, and in this contention I am sure

THE NATIONALIST PERIOD

I shall have the support of the many British composers whom it has been my privilege (thanks to the broadminded policy of the Corporation of Bournemouth) to assist.

Sir Dan Godfrey's letter errs on the side of overstatement, for Birmingham was then only at the beginning of its work in municipal orchestral music, and the other towns he mentions had not justified his claims for them. Had he confined his remarks to the progress made in his own town, Bournemouth, Sir Dan's case against Sir Thomas would have been stronger. But it is not in the nature of pioneers to suffer calmly any attack on their efforts.

THE TWENTIETH CENTURY ORCHESTRAS

AMONG many interesting persons with whom Sir Dan Godfrey came in contact in the course of his musical work was the Kaiser Wilhelm II. At a time when the Imperial yacht was anchored in the Solent, the Kaiser offered the services of his band for a concert in Bournemouth, and Godfrey had a rough passage in a small launch when going out to make the necessary arrangements on behalf of the Corporation. The German Emperor had come on a mission of goodwill, and it is not to be charged against him that he offered the services of his private band instead of attending in person the Bournemouth concerts: he honestly felt that he could best serve the cause of Anglo-German friendship by taking the course he did.

Anglo-German friendship was not so much a private wish as a manifestation of political policy. There was at that time much uneasiness in Britain about the use to which the rapidly-growing German fleet was to be put, and convincing demonstrations of friendship were necessary to gain time for the maturing of German plans. German music was that country's most acceptable ambassador, offering a most favourable impression of the German people to the Western Powers, and especially to Britain. The German people, however, who for long had been led to think of European nations to the east as their cultural inferiors, were now, for reasons best known to the German High Command, beginning to hold similar opinions about nations to the west. Since there could be no doubt about the superiority of nineteenth-century German music over its contemporary British cousin, there was a growing contempt for *das Land ohne Musik* about the time when Wilhelm II came to the throne. It stung British musicians to reaction in favour of British music.

There had been murmurings about the neglect of British conductors when Richter took control of the Birmingham Triennial

THE NATIONALIST PERIOD

Festival in 1885, but the arguments were petty, and only those whose interests were at stake pursued them with any ardour. Far more decisive was the resistance to the German contempt when it became apparent. Of some importance in the 'nineties was the German Club in London; there one could meet musical celebrities and hear intelligent criticism of modern music. One night in 1895 Carl Armbruster was holding forth. The gist of his remarks was that only a German could conduct Wagner.

In the room was Dr. George C. Cathcart, an eminent laryngologist.

'I know one Englishman who can conduct Wagner', he said.

The remark was greeted with jeers. 'An Englishman conduct Wagner! Rubbish!'

The jeers so rankled in Cathcart's mind that he left the room and decided on a course of action that he might not otherwise have taken. He sought out Robert Newman, who was trying to raise the capital for a new series of promenade concerts at Queen's Hall, and offered an investment on the conditions that French pitch was used. The conductor already appointed was Henry J. Wood. That decision cost Cathcart £2000, for all but two concerts of the first season resulted in financial loss. The Queen's Hall Promenade Concerts, however, were established, and Henry J. Wood entered upon his long career as their conductor.

Cathcart's advocacy of low pitch came as a result of his researches into the causes of voice-strain. Handel's English tuning-fork stands at A422.5, which is almost the same as the pitch first adopted by the Philharmonic Society in 1813 — A423.7. By Costa's time, however, the pitch in use at this Society's concerts was A452.5 — a difference of over a semitone above the classical pitch.[1] Under Wood's direction a pitch of A439 at a temperature of 68 degrees Fahrenheit was adopted, and the organ at Queen's Hall tuned to this pitch meant that it would have to be adopted by anyone wanting to use that

[1] The highest pitch used in England was that of the Strauss Band in the Imperial Institute, London, in 1897, which was A457.5.

THE TWENTIETH CENTURY ORCHESTRAS

organ. Fortunately the move got more support than that, for the programmes of the first Queen's Hall Promenade Concert bore a note: 'At these concerts the French Pitch (Diapason Normal) will be exclusively used. Mr. Newman is glad to say that it will also be adopted in future by the Philharmonic Society, the Bach Choir, the London Symphony, Mottl, and Nikisch concerts, and in concerts under his direction, which begin on October 6th.' A439 was named the New Philharmonic Pitch to distinguish it from Costa's A452.5, which was called Old Philharmonic Pitch.

Prices of admission to the Queen's Hall Promenade Concerts were fixed at one shilling for the promenade, two shillings for the balcony, three shillings and five shillings for the grand circle. Newman had decided to keep to traditional Promenade Concert prices, having been previously familiar with these at certain Covent Garden Theatre Promenade Concerts conducted by Sir Arthur Sullivan and Rivière — that same whom Henry Wood went to study in Llandudno. Covent Garden Promenade Concerts had not been a financial success, but Newman determined to try again in his new hall. The Queen's Hall Promenade Concerts were to be free and easy, with stalls set round the auditorium from which various commodities could be bought. And, of course, a promenading audience.

The programme of the first Queen's Hall Promenade Concert must be seen to be believed; this is what Newman's customers got for their shillings, three instrumental soloists and five singers, in a list of twenty-one musical items:

PROGRAMME

Overture	Rienzi	*Wagner*
Song	Prologue: Pagliacci	*Leoncavallo*
	Mr. Ffrangçon Davies	
(a)	Havanera	*Chabrier*
(b)	Polonaise in A	*Chopin*
	Orchestrated by Glazounov	
Song	Swiss Song	*Eckert*
	Madam Marie Duma	
Flute Solos (a)	Idylle	*Benjamin Goddard*
(b)	Valse from Suite	

THE NATIONALIST PERIOD

Song	Thou hast Come	*Kennington*
	Mr. Ivor McKay	
Chromatic Concert Valses from the opera Eulenspiegel		*Cyril Kistler*
	(First performance in England)	
Song	My Heart Thy Sweet Voice	*Saint-Saëns*
	Mrs. Van der Vere Green	
Gavotte from Mignon		*Ambroise Thomas*
Song	Vulcan's Song (Philemon and Baucis)	*Gounod*
	Mr. W. A. Peterkin	
Hungarian Rhapsody in D min. and C maj. (No 2)		*Liszt*

INTERVAL 15 MINUTES

Grand Selection	Carmen	*Bizet*
	(arr. Cellier)	
Song	Largo al Factotum	*Rossini*
	Mr. Ffrangçon Davies	
Overture	Mignon	*Ambroise Thomas*
Cornet Solo	Serenade	*Schubert*
	Mr. Howard Reynolds	
Song	My Mother Bids Me Bind My Hair	*Haydn*
	Madam Marie Duma	
Bassoon Solo	Lucy Long	
	Mr. E. F. James	
Song	Dear Heart	*Tito Mattei*
	Mr. Ivor McKay	
The Uhlan's Call		*Eilenberg*
Song	Loch Lomond	*Old Scottish*
	Mrs. Van der Vere Green	
Song	The Soldier's Song	*Mascheroni*
	Mr. W. A. Peterkin	
Valse	Amoretten Tanze	*Gungl*
Grand March	Les Enfants de la Garde	*Schloesser*
	(First Performance)	

Wagner in 1855 had complained of the impossibility of rehearsing long programmes at one sitting, but the Philharmonic Society had never asked him to perform such double-length programmes six nights a week for ten of the hottest weeks of late summer. Yet this is what Newman required of Henry Wood and his men. This was a form of artistic gluttony that had its origin in the long programmes of the Victorian music-hall. Time has mellowed the taste of the Prom. programmes. Such programmes harked back to the days of Jullien the charlatan, without his redeeming virtues. Indeed the influence of Jullien is to be seen in many of Wood's early programmes. Friday, September 6th, 1895, is a case in point. It is described as a Military Night:

THE TWENTIETH CENTURY ORCHESTRAS
PROGRAMME

Military March — orchestrated by Harold Vicars		*Schubert*
Overture	Les Dragons de Villars	*Maillart*
Aria	Valse Song (Romeo and Juliet)	*Gounod*
The German Patrol		*Eilenberg*
Recit. and Air	Sound an Alarm	*Handel*
Military Overture in C		*Mendelssohn*
Song	The Soldier's Tear	*Alex Lee*
Recitation	The Defence of Lucknow	*Tennyson*
Selection	The Red Hussar	
INTERVAL 15 MINUTES		
The British Army Quadrilles		*Jullien*
Song	The Castilian Maid	*Lisa Lehmann*
Cornet Solo	Sweet Sixteen	
Song	The Temple of Light	*J. Valentine Hall*
Drum Polka	.	*Jullien*
Song	Jeannette and Jeannot	*Chas. W. Glover*
The Soldiers' Chorus from Faust		*Gounod*

Newman's 'enterprise', in fact, was reactionary to the last degree. Nor can it be justified on business grounds, for it ignored factors constant in the demand of London audiences. One of these was a liking for one work of serious symphonic dimensions in contrast with the light fare, and the other was a liking for novelties which the Military programme did not consider. These factors were well known to Jullien, and Manns had been demonstrating them to Crystal Palace shareholders for forty years with satisfactory returns. Dan Godfrey had found the same characteristics present in Bournemouth audiences, and by 1895 had established symphony concerts as a primary attraction. Queen's Hall Promenade Concerts became profitable only when Wood adopted the same tactics. To Newman's credit it must be said that he was open-minded, and gave Wood every support in bringing about the changes that have made the Promenade Concerts the greatest single factor in London's musical life. But the credit for the ideas must go mainly to Wood.

Besides artistic changes, there were changes of policy of some importance inside the orchestra. Chief of these was the rejection of a long-established deputy system. Throughout the history of the orchestra in England a *laissez-faire* policy had predominated. Casual employment of players was the rule;

THE NATIONALIST PERIOD

the nearest approach to stability of employment being seasonal engagements by contract, and as was seen in 1860, players had even incurred the displeasure of so important an organization as the Philharmonic Society by honouring these contracts. But players wanted to keep on good terms with all their numerous prospective employers, for even such contracts as were available were for short periods. They therefore reserved the right to send deputies when unable to keep an engagement through illness or in the event of being asked to appear at some other concert promoted by an organization with whom the player may have been on satisfactory business terms for years. It was one of the evils of *laissez-faire*, but one that operated against the employer instead of, as usual, against the worker. The first season of the Queen's Hall Promenade Concerts had been going on for seven weeks when the results of the deputy system came home to their conductor:

> I realized something of a nightmare on the morning of September 30th. I found an orchestra with seventy or eighty unknown faces in it. Even my leader was missing. Arthur Payne, the deputy leader, told me of a certain musical festival. My regular players were all there. Moreover, they would be absent for a week.
> I had to put up with this sort of thing for years. It was hardly fair for a young conductor to have to rehearse a week's concerts in three mornings (nine hours all told) with new and often inexperienced players. I made up my mind there and then to fight what is a bad principle, but little did I think how many years it would take me to bring about a reform. The public was none the wiser because it had not yet learned to distinguish individual players. Neither could the papers tell them anything because all the critics were at the festival.
> One result it did have. Fry Parker never led the Promenade Concerts again. Arthur Payne took his place.[1]

Not until 1904 was Henry J. Wood able to overcome the deputy system at the Proms. By that time he had established

[1] Sir Henry J. Wood, *My Life of Music*.

THE TWENTIETH CENTURY ORCHESTRAS

a reputation equal to Costa's as a martinet. Players knew that late arrival was an unforgivable offence, and that popping out of the room for a quick smoke during a tacet movement was dangerous. (Newman used to go round and lock the doors.) If a singer, no matter how famous, failed to come up to Wood's expectations at rehearsal, he might have to listen to a reminder about the elementary principles of voice-production. Still the deputy system went on. One morning Newman walked on to the platform during a rehearsal. In his hand he carried a paper. He read: 'Gentlemen. In future there will be no deputies. Good morning.' And he strode off.

Nothing was said, but next morning the resignations of forty players were received.

Only by the most careful planning was it possible to get through the enormous programme of music required for any season of the Proms. Everything had to be thought out months ahead, so that when the time came for rehearsal every second would be allocated for its purpose. It did not take a new player long to discover that everything went by clockwork — the clock being a large old-fashioned watch on the conductor's desk. Major irritations were the result of the timetable being disarranged, but this the conductor never did; generally the reason would be found in incorrect parts or some composer floundering in control. True, the timetable was so strict that a composer had little chance to make last-minute alterations: Sir Dan Godfrey's system gave more freedom to the young composer who had to rely as a last resort on trial and error, but no composition was ever allowed to appear in performance in a form confusing to the players. Havergal Brian's first experience at the Proms. is a case in point:

> One morning I received a letter intimating that Mr. Henry Wood had put down two of my works for performance at his 1907 Promenade Concerts, *English Suite* and *For Valour*. I was warned to have the parts correct and sent to the librarian of the Queen's Hall Orchestra one week before the concert. It all seemed so fine and promising; the end

THE NATIONALIST PERIOD

of long striving, and I daresay I was proud of my achievement. But I got a rude shock. On returning from a day in the country one Saturday afternoon I found a telegram awaiting me. It said: 'First rehearsal of your English Suite this morning. Parts all wrong. Must be correct by 10.30 Monday morning's rehearsal.—Henry J. Wood.' Only a few months before I had conducted this same work in Leeds Town Hall with the Municipal Orchestra before a crowded audience, so I became more and more mystified by Wood's telegram. I caught that evening's train to Euston, and on Sunday morning made my way to Queen's Hall. There I was introduced to Mr. W. Tabb, the then orchestral librarian, who very kindly pointed out that I had not given the brass players a single cue. After the rehearsal Wood said he was sure the work would be a great success, and he showed his faith in it by giving it two more rehearsals.

This was a mild disturbance compared with some. It speaks well for Sir Henry's organization that with the enormous amount of work to be done he could yet make time for three rehearsals of one work. The situation described by Havergal Brian could not have developed in later years, however, for Wood finally resorted to correcting all music himself. Every player saw on the top of his copy the words 'Corrected. Henry J. Wood' and the date. In the earlier concerts every device would be tried to reserve rehearsal time. Sometimes a solo pianist would elect to play a sonata instead of a concerto; this would ease the situation, and prevent resource to that last ditch of all — concentration on the difficult works at rehearsal and relying absolutely on the stick at the concert for the performance of familiar works. The stick, however, was quite reliable; no player could go wrong unless he ignored it. Sir Henry used to tell his players that he put in half an hour's practice with the baton every morning, and nobody doubted it, for thoroughness was of the essence of Sir Henry. His tuning parade, when every player filed past him and tuned his instrument under the conductor's eye was but one of the ways in which Sir Henry's thoroughness was manifested.

THE TWENTIETH CENTURY ORCHESTRAS

A vital factor in the success of the Queen's Hall Promenade Concerts, however, was the audience. It is an audience with a distinct personality. The audiences at Jullien's Promenade Concerts had been workers and visitors from the country who sought popular entertainment free from the vulgarity that prevailed in the theatres as they then were. The Queen's Hall audience was not of that type, although it was often insisted that they were quite ordinary people. There are, however, some features that mark promenaders out for distinction. Ignoring the collarless, be-sandalled, be-bearded men and the long-haired cape-draped women who seem to drift into any artistic gathering in London, there is a good proportion of students who find that in a ten weeks' course of concerts they can refresh their knowledge of practically the whole range of standard orchestral works, and hear in addition many new works by which they can form opinions on the trend of the times. This they can do at a price students can afford. Such people have good reasons for regular attendance at the Proms., while the casual visitor can get a good general impression of a favourite composer by attending one or more of the single composer evenings for which the Proms. are noted. The out-and-out promenader, however, who boasts of never having missed a Prom. in years, is a queer sort of intellectual snob. He affects to despise those who sit in the gallery or circle, holding a perverted opinion that the physical discomfort of standing through a long programme on a hot evening is necessary for the enjoyment of music. To his credit it must be said that his snobbery is a good antidote to that of Covent Garden's opera patrons, where money talks. Neither is ideal: if Covent Garden preserves the custom of caste, Queen's Hall preserves the Victorian custom of swooning. The physical strain on habitual promenaders is of their own choice, however, but that put on the players is a hardship: long programmes day after day, with an efficient conductor getting the last ounce out of his men the whole time. The first half of a promenade concert itself was until recently equal in length to a full symphony concert. The

THE NATIONALIST PERIOD

strain begins to tell on the players towards the end of the season; there is considerable physical exhaustion in some sections of the orchestra, and throughout the band nerves are generally on edge. Towards the end of the season the standard of performance was liable to fall off; visitors familiar with Sir Henry Wood's interpretations at provincial festivals and concerts, coming to London for a Prom., have been known to form a poor opinion of London orchestral playing. In justice to the players, attention should be called to Mr. Bernard Shore's experience as set down in his revealing book *The Orchestra Speaks*. Shore says, 'the exhaustion experienced after a Promenade concert becomes nearly intolerable as the weeks go on'.

Yet the amount of work done by any player was far less than that done by Sir Henry himself; and this amazing man kept it up for fifty years! The Proms., moreover, were but a small part of his work. In Manchester he conducted the Brand Lane Concerts and the Gentlemen's Concerts; he conducted the Liverpool Philharmonic Society for a time, and concerts in Nottingham, Leicester, Hull, Wolverhampton, etc., in addition to festivals at Sheffield, Norwich, Birmingham, Westmorland, and other places.

Throughout the nineteenth century the orchestras at provincial festivals had been amateur stiffened with professional players. They came together for the festival only, but there was no general recognition of the inefficiency that is inseparable from such scratch orchestras. In most cases a conductor of national repute was engaged, but there were exceptions — notably at the oldest festival of all: the Three Choirs. Here, although in the early days of these festivals William Boyce had been brought from London to direct the performances, tradition had grown up for the organist of the cathedral concerned to take charge. Scratch orchestras remained the rule at Three Choirs Festivals until after the Four Years War, when the London Symphony Orchestra was engaged *en bloc*.

The London Symphony Orchestra came into being in 1904,

THE TWENTIETH CENTURY ORCHESTRAS

from a nucleus formed of the forty players who seceded from the Queen's Hall Orchestra as a result of Wood's objection to the deputy system. These players combined to form an orchestra with a new outlook. Theirs was a co-operative venture; the players jointly took the financial risk and controlled the orchestra's policy. One result was that instead of the conductor giving employment to orchestral players, the orchestra engaged its conductor. This was a further step along the road that started when Banister left the king's service and established his commercial music-meetings in the seventeenth century. Freedom is a desirable thing, but it brings responsibilities. In the eighteenth century there were musicians like Handel, J. C. Bach, and Salomon who accepted this responsibility, and Sir Charles Hallé for a time did the same thing in the nineteenth century, but by this time concert promotion had become a big business, and the best results were obtained when Hallé had the co-operation of an influential local committee. Business practice has been for long in favour of the limited liability company as the most suitable way of financing even moderately-sized enterprises, and while this system had no attraction for the Philharmonic Society of London or the Hallé Society of Manchester — neither of which existed for profit-making — there was bound to come a time when the advantages of a widely-spread capital responsibility would appeal to players whose livelihood depended on the promotion of orchestral concerts. The players who formed the London Symphony Orchestra founded a self-governing orchestra, and their lead was followed in the following year, 1905, by three players — John Saunders, violinist; Eli Hudson, flautist; and Charles Draper, clarinettist — who founded the New Symphony Orchestra, which was incorporated in 1907 as a limited liability company and has become well known since 1920 as the Royal Albert Hall Orchestra. Thus did modern methods of finance come into orchestral organization.[1]

---

[1] Such companies as the Crystal Palace Company carried on orchestral concerts as part of a larger business.

THE NATIONALIST PERIOD

It goes almost without saying that the conductor to whom the London Symphony Orchestra first turned was Hans Richter. For so long players had been obliged to bear with the uncertain tempers of difficult conductors that the choice of one who was a complete master of his craft, and respected his players as colleagues, was to be expected. These men had freed themselves from economic dictatorship, but they themselves demanded strict discipline in their artistic work. It will be recalled that in 1860 players had similarly stuck to Costa, the martinet, because he never abused his position by siding with the management against the economic interests of his men, but, nevertheless, drilled them in performances until their reputation as an orchestra stood above that of their rivals. The London Symphony Orchestra arranged for a series of concerts to be given under Richter in London, the first of which took place on June 9th, 1904, and other conductors of the highest merit followed — Nikisch, Steinbach, Colonne, Elgar, and Sir George Henschel. Under the influence of such men as these the London Symphony Orchestra soon began to develop features that were to make it famous. The first of these was an extraordinary adaptability to strange conductors, always a feature of London orchestras, but now more marked than ever, and the second a nobility in their string tone that may have had something to do with Elgar, but was more likely due to the training of Nikisch, himself an excellent violinist, and one who claimed that every conductor should learn string-playing not only because it made him absolutely familiar with the foundation of the orchestra, but gave a flexibility to the wrist essential for a distinct beat. Richter, however, had most influence on the London Symphony Orchestra in its first years, and no doubt would have continued to be their favourite had he remained in this country. Elgar had a warm regard for their leader, W. H. Reed; regard not only as a musician, but as a man. A tour of the provinces took place in 1905, conducted by Elgar, followed by a visit to Paris, where the orchestra played under Colonne, Messager, and Stanford. 1906 saw the

THE TWENTIETH CENTURY ORCHESTRAS

London Symphony Orchestra at Covent Garden for a special season of German opera conducted by Nikisch, Reichwein, Schalk, and Ysaye. Safonoff also conducted them in concert work.

1909 saw the introduction of Elgar's *Symphony in A flat* to an expectant public. By this time everything he wrote was being enthusiastically received. The symphony was first played in Manchester by Richter and the Hallé Orchestra, and four days later was given in London by the London Symphony Orchestra with the same conductor. By this time this orchestra's reputation stood as high with the public as any orchestra in London; they 'discovered' Koussevitsky, who came to London as a virtuoso double-bass player; with them he conducted a Beethoven programme in 1909, and brought an authentic interpretation to music of the Russian nationalist school, which was then gaining favour in Britain. Thomas Beecham conducted the London Symphony Orchestra first in the 1910-11 season, and in the following year the orchestra undertook a tour of the U.S.A. and Canada, conducted throughout by Nikisch.

1912 is a convenient year in which to pause and take stock of the general situation in English orchestral circles, for in this year the Philharmonic Society of London celebrated its centenary, Myles Birket Foster published its history, and in recognition of the Society's invaluable work for music in England His Majesty the King graciously permitted it to be called henceforth The Royal Philharmonic Society. During the latter half of its hundred years of existence the Society had recovered from the ultra-conservative trend that had been creeping in when Hogarth ended his history of the first fifty years of the Society's existence, but the Society no longer had the same virtual monopoly of new compositions that the conditions of 1813 had permitted. Now, in 1912, there were the Queen's Hall Promenade Concerts giving an amazing number of first performances each year, and Thomas Beecham with ideas that were nothing if not original. Nevertheless, it was the old 'Phil' that had pulled British orchestral music through the doldrums of

THE NATIONALIST PERIOD

the later nineteenth century, by reason of its peculiar constitution as a sort of professional club and non-profit-making society. Personal friendships with Macfarren, Sullivan, Mackenzie, Parry, and Stanford gained approval for their compositions at a time when the German shop-window held an unrivalled display. 1889 saw Stanford's *Violin Suite* performed by Joachim, and Parry's *English Symphony* at the Phil., and there was plenty of music by Macfarren and Sullivan to be heard about this time. Mackenzie's *Twelfth Night* came in 1890. Frederick H. Cowen, the Society's conductor, complained that the number of composers who elected to conduct their own works had an unsettling effect on the orchestra, which was probably true, but there was no diminution of first performances. Hamish McCunn's overture *The Ship o' the Fiend* appeared in 1896, and the following year brought forth at Philharmonic concerts Mackenzie's *Scottish Concerto* for pianoforte, Stanford's *Pianoforte Concerto in G*, McCunn's *Highland Memories* and Parry's *Symphonic Variations*. It was in this year that Elgar offered them the first performance of a new orchestral work, but would not let them see it before acceptance.

All these British works were of little merit, however, in comparison with those from abroad. Tschaikovski's symphonies began to appear in the 'nineties: No. 4 in 1893, the *Pathetic* in 1894 — this was its first performance in England, and it was received with such favour that a repeat performance had to be given at the next meeting — Tschaikovski's *Fifth Symphony* came to London the following year, not through the Philharmonic Society, however, but at one of Nikisch's concerts. 1897 and 1898 saw performances of the first and third Symphonies of Brahms at Philharmonic concerts, and 1899 a repetition of his fourth, which the Society had previously heard in 1887, but the outstanding event of this year was a visit of Richard Strauss, who conducted *Tod und Verklärung*. The next decade was to see Strauss in the ascendant, a centre of controversy and an acknowledged master of what was then regarded very generally as the orchestra of the future. In contrast to

THE TWENTIETH CENTURY ORCHESTRAS

Strauss stood Debussy, but Strauss was the composer better known in orchestral circles. His large orchestra, with its pedal clarinet, clarinet in D, and, even more remarkable, devices that had up to that time been confined to the 'noises off' at a melodrama, gave the man in the street more to talk about than the aesthetic innovations of Debussy.

Graphs showing the number of new performances given by the Philharmonic Society over ninety years, arranged according to nationality, will be found on page 258, but figures are inadequate to describe artistic trends, since they indicate neither the relative importance of compositions nor the measure of public response. Richter's production of Elgar's *Enigma Variations* in 1899, and his *Symphony in A flat* in 1909, far outweighed the whole of the Philharmonic Society's offerings of British music during that period; but the Philharmonic took up the challenge in 1910 with the first performance of Elgar's *Violin Concerto*. Kreisler played the solo part. So many people had to be turned away at the first performance that a second performance had to be arranged. Again the hall was found insufficiently large to accommodate all who wanted to hear Elgar's concerto.[1]

Side by side with this fondness for Elgar went a vein of criticism of native effort annoying to its recipients but quite in keeping with the British way of life. In the House of Commons a healthy opposition is supposed to be welcome, and one expects it; but in the shifting sands of musical criticism any sort of criticism may come from any quarter. Sir Thomas Beecham has observed that although the British public had placed Elgar on a pedestal higher than that of any composer since Purcell, he did not find this valuation shared by either our own or foreign musicians. Such statements are valuable aids for keeping majority opinions up to par, but minorities are by no means immune from them; Sir Thomas has had his share of criticism,

---

[1] During that same year, 1910, Elgar's *First Symphony* had over a hundred performances. It would be interesting to know if any other composer in any country at any time has met with such appreciation.

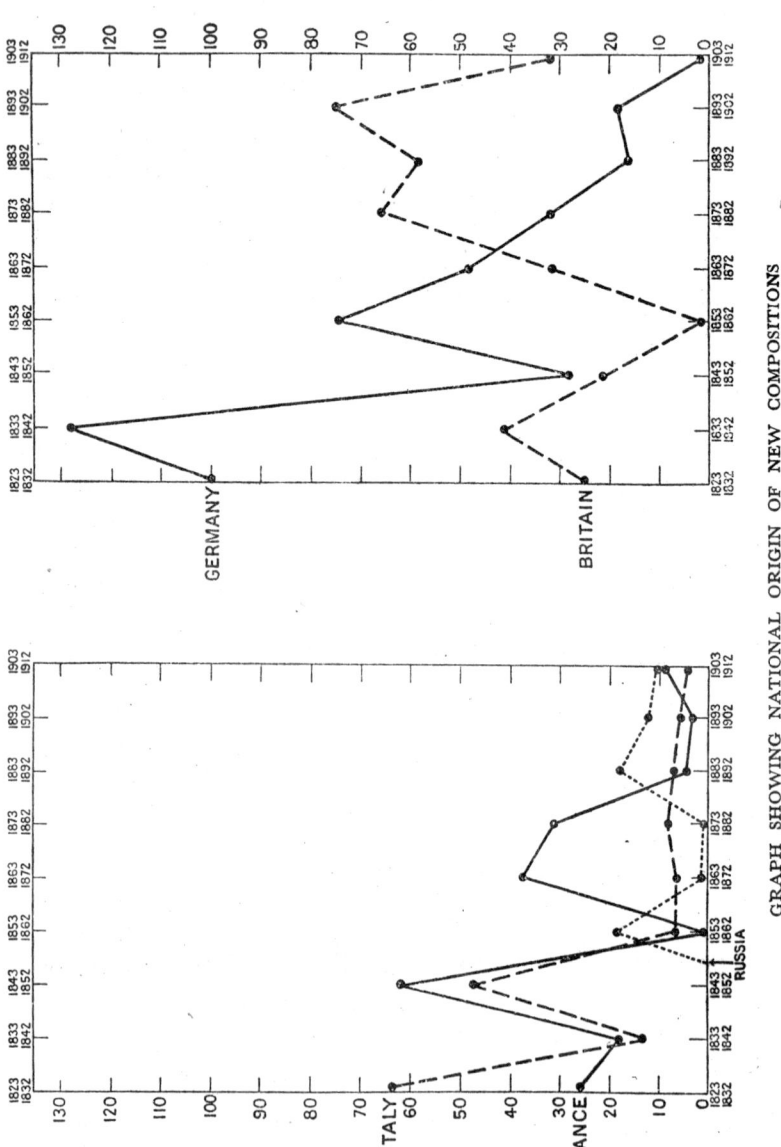

GRAPH SHOWING NATIONAL ORIGIN OF NEW COMPOSITIONS PLAYED BY THE PHILHARMONIC SOCIETY OF LONDON. 1823 TO 1912

THE TWENTIETH CENTURY ORCHESTRAS

some of it invited, some of it prejudiced. Of the latter must be mentioned the tardy acceptance by British opinion of his triumph with a British orchestra in Berlin in 1912.

All things considered, orchestral playing must have reached an exceptionally high standard under Beecham in that year, for the Berliners took a keen interest in the Beecham Orchestra's performances, although they could not unreservedly accept Beecham's interpretations. 'It sounds grand, but it isn't Mozart', said an eminent professor. 'They wondered', says Sir Thomas, 'at the technical accomplishment of the orchestra, notably the wind and horns, which were without question superior to their own. One of the leading critics devoted an entire article to a close analysis of the style, tone, and method of my players, indicating in detail wherein they differed from those of the Austrian as well as the German orchestras, and ended by saying, "These Englishmen play with a sovereign authority all too rare nowadays anywhere".' Naturally, the players expected to be warmly welcomed on their return to London, but little public interest was shown in the event, and a newspaper critic actually minimized the German reports. 'I thought', writes Sir Thomas, 'of the almost overwhelming fuss made by the people of every other country when one of their representative institutions ventured across the frontier for a week or even a few days. We had been in the stronghold, indeed in the utmost citadel, of the world's music for over two months, had conquered its prejudices, won its suffrages, and had been admitted to a status held by only two or three of its own organizations. This, however, in the England of 1912 passed almost unnoticed, and, for some reason I was never able to understand, was even a little resented.'

Standing before the German shop-window, warding off the missiles of honourable democratic opposition, Sir Thomas might well be puzzled. There is no logic in this sort of thing; had the Germans derided British orchestral music, as Hans Richter did after his return to Germany, the opposition would have supported the national cause with all the force at its

THE NATIONALIST PERIOD

command. Nobody knows this better than Sir Thomas, but like all artists he is sensitive to criticism. One must indeed be thankful for his sensitivity, for if it prejudices his opinions at times, it is the secret of his amazing genius in the interpretation of music. No man has been better able to explain this than Sir Thomas himself. In his autobiography he writes thus of his first London concerts:

> My opening essay was given at the Bechstein (now Wigmore) Hall, with a body of forty players drawn from the Queen's Hall Orchestra, and the programme included several of the eighteenth-century French and Italian works which I had collected on the Continent. A qualifying note of modernity was Cyril Scott's pretty ballad for voice and orchestra, *Helen of Kirkconnel*, sung by Frederick Austin, who not long before this had thrown up his old job in Liverpool to devote himself wholly to the profession of singer. My chief sensation was one of definite disappointment with myself: for at hardly any moment during it had I the conviction that I was obtaining from my executants the tone, style, and general effect I wanted. Somehow or other the sound of much of the music was strangely different from the conception of it in my brain, and though my friends did their best to make me think I was mistaken, and the newspapers were sympathetic, I felt I knew better, and I knew I could do better. But first I must find out what was the matter. I returned to the study of a large number of well-known scores, attended during the next two months nearly every concert at Queen's Hall, and found that in many instances I experienced the same sense of dissatisfaction on listening to performances under other conductors. Years before I had not been troubled in this way; everything had sounded grand and perfect, and I began to be alarmed. Was my ear beginning to be affected or — more awful reflection — had the actual sound of the modern orchestra begun to distress me as it did an ultra-fastidious friend of mine in Paris? But one evening I listened to a highly unsatisfactory rendering of some favourite piece, of which I knew every note, and now there could be no doubt

where the fault lay. At one moment the brass instruments were excessive, at another inadequate; the horns strident or feeble, and the strings feverish or flaccid. Briefly, there was no true balance or adjustment of the component parts of the machine, and it began to filter through my consciousness that if here was the source of trouble in a flagrant instance like this, it might turn out to be the same in fifty others less obvious. My curiosity well aroused, I followed with a keener ear everything I heard, and formed a conviction which the passage of time has only strengthened. The supremely important factor in any choral or instrumental ensemble is the relationship between the different sections of the forces of play.

Throughout his career this supersensitive feeling for nicety of balance has led Beecham on to create original readings of the orchestral works of the old masters, and to bring to life the beauties of modern works. 'It sounds grand, but it isn't Mozart', may be a just criticism from the point of view of the critic, but to Sir Thomas Beecham there is only one true reading, and that is arrived at as a result of a keen ear and a prolonged study of the score. There is no more remarkable instance of this than Beecham's persistent championing of Delius, the beauties of whose orchestration he was able to appreciate well in advance of other conductors.

Like Sir Thomas, Delius was an individualist with a fondness for the stimulation of self-justification among audiences. An eminent authority assured him in 1907 that only two London orchestras were available, Queen's Hall and the London Symphony Orchestra, and yet he found Beecham conducting a third, playing modern music with relish. 'London is the only town in the world where a first-class band like this can give such a set of concerts without one of its leading musicians being aware of its existence', he said, speaking no doubt from a complete knowledge of the opinions of them all gained during the previous week. Fortunately Delius and Beecham understood and were equally amused by London opinion; both were

THE NATIONALIST PERIOD

northerners and unlikely to take seriously the convention that metropolitan opinion was all-important. Bradford, it is true, offered little to Delius, and Liverpool offered little scope for Beecham's musical activities, but both places had had their effect on the boyhood of the men.

Beecham, the son of a wealthy manufacturer in St. Helens, had been familiar with Liverpool musical life in his boyhood. The city was considerably more class-conscious than Manchester or Bradford, with a social system that clung to the exclusive type of subscription concert that had played such an important part in London's musical life during the eighteenth century and the first half of the nineteenth. Hallé had broken down this system in Manchester, but had never had sufficient authority in Liverpool to do so. Right into the twentieth century a statement used to appear on the programmes of the Liverpool Philharmonic Society's concerts:

> No gentleman above twenty-one years of age residing or carrying on business in Liverpool, or within ten miles thereof, and not being an officer of the Army or Navy or Minister of Religion, is admissible to the Boxes or Stalls at the Philharmonic Society's Concerts, unless he be a Proprietor, a member of the Family residing at the house of a Proprietor, or has his name upon the list of Gentlemen having the *entrée* exhibited in the corridors.

The gallery was open to all, but only after a good deal of formality could one get into other parts of the building. And yet in a city so socially backward there was a progressive policy in music. Although Bantock's experiment at New Brighton was arbitrarily terminated, he continued his association with Liverpool as conductor of the Liverpool Orchestral Society after he had gone to live in Birmingham. The founder of this society, Alfred E. Rodewald, had some points in common with the Delius family at Bradford. He was a cotton merchant of German origin but English outlook, having been educated at Charterhouse. For the programmes of a memorial concert held

THE TWENTIETH CENTURY ORCHESTRAS

in 1903 Ernest Newman, himself a Liverpool man, supplied the following note; he says that in 1884 Rodewald

> took the first step along that path on which he was destined to do so much for music in Liverpool. In that year he founded an amateur orchestral organization to assist at the Saturday Evening Concerts of Father Nugent: its title was 'The People's Orchestral Society'. Six years later it was reconstituted as 'The Liverpool Orchestral Society'. At first it was largely amateur in its constitution, but as time went on, and ambitions grew, and the need was felt for a band that could cope with the most difficult music of modern days, the professional players gradually came to outnumber the amateurs. This fine body of performers Mr. Rodewald infected with his own enthusiasm ... He was an amateur only in the dignified and honourable sense that he gave his services freely and cheerfully to his beloved art. He was emphatically no amateur in any other sense. Most professional musicians might well envy the range, the lucidity, the accuracy of his knowledge of the modern orchestra. As a practical conductor, too, he steadily improved year by year ... He was, in short, a thoroughly accomplished musician, and if he was appreciated while he was alive, it is safe to say that the musical public of Liverpool will in the future realize still more clearly how vast were his services to music here. For it would be a mistake to suppose that those services began and ended with his conducting of these concerts ... He was always ready to assist the Sunday Society to place high class music before the people, and when Mr. Granville Bantock left New Brighton, Mr. Rodewald took up the good work there and did his best to keep the sacred flame of art alive in the community.

Sir Thomas Beecham has stated that it was the ambition of every young musician in Liverpool to join Rodewald's orchestra, and that this ambition stimulated an interest in all kinds of orchestral instruments. This could not have had any permanent results, however, had there been no opportunities for expert

THE NATIONALIST PERIOD

tuition, and there was not, as in Manchester, a college of music at which players could be trained. Manchester was near enough for many people, however, but not the poorer amateurs; fortunately another avenue was open to them. This came as a result of Liverpool's love of opera.

Again we must revert to the Liverpool Philharmonic Society. Picton wrote of the Liverpool Philharmonic in 1873: 'Although the performances are not operatic, the *morceaux* selected from the operas being presented in a fragmentary manner, without scenery or dramatic effect, the institution discharges for Liverpool society the same functions as Her Majesty's Theatre does for ... the Metropolis. It is here that the small talk, persiflage, and gossip of what is called "good society" pervade the air and circulate their agreeable flavour ... The music however is not all operatic; oratorios, masses, cantatas and choral music of all kinds vary the bill of fare, which is generally selected with taste and performed with skill.' Alas, the wish for an operatic season had to be met, and met it was by a solid support of the Carl Rosa Opera Company, so solid that Carl Rosa bought the biggest theatre in Liverpool and made it his headquarters. From Christmas until spring each year the Carl Rosa opera season ran in Liverpool to crowded houses; not to be seen there regularly was a sign of civic unworthiness, and the Carl Rosa Opera Company supplied the demand for teachers of orchestral instruments during their stay.

Like Rodewald's orchestra, the Carl Rosa Opera Company was not entirely professional. It was their system to tour the country with a good orchestra having some professional strings and all essential soloists in the wood-wind and brass departments first-rate players, mostly foreigners; the remainder of the strings were recruited in the towns they visited, and, in the week before the company was due to appear, rehearsals of the local players took place under the Carl Rosa conductors (the company employed two), who travelled daily to the town for this purpose. Contact was therefore made between the Carl Rosa players and the amateurs in the towns they visited, and

THE TWENTIETH CENTURY ORCHESTRAS

many amateurs took advantage of the presence of Carl Rosa in their own and nearby towns to take lessons from the orchestral players. Thus the opera had a beneficial effect on Liverpool's musical life apart from its own performances. When Thomas Beecham was sent away to school he missed Rodewald's concerts, but by the time he had passed through Oxford and travelled abroad he had outgrown his love for the travelling opera company and the semi-amateur orchestra. In later life he strove always for the highest technical proficiency in the players he engaged, but it is of interest to know that his first engagement as an opera conductor was with a very second-rate repertory company, and was obtained as a result of his ability to play popular operas from memory. His experiences with this company only hardened his conviction that opera with such companies was not only sordid but ridiculous. He believed in the power of the wealthy patron, and distrusted the opinions of those who appealed for good music on a broad democratic basis. If in this he came to conclusions opposite to those obtaining in Liverpool, Manchester, the Crystal Palace, and Queen's Hall, he did so with a complete knowledge of the facts.

Beecham became associated with the New Symphony Orchestra in 1906 and sought to introduce this orchestra into the private salons of London society. Because of its small size and refined style this was practicable, and there was scope for the performance of novelties at private concerts. He might even start a fashion for orchestral concerts in Mayfair much as Hallé started a fashion for afternoon Beethoven recitals in the 'fifties. True, hostesses who would gladly spend a thousand pounds on a galaxy of star soloists had to be persuaded to consider an orchestra; but it was a novelty, and hostesses had to be alive to novelties if they were to be leaders of fashion.

Such hostesses lend themselves easily to ridicule in print. Hallé's stories have a modern equivalent in the second chapter of Aldous Huxley's novel, *Point Counter Point*. But the Edwardian period was among the richest in English social life, and the

THE NATIONALIST PERIOD

patronage of the upper middle classes bore the financial burden of the art of most of their critics. This was the great era of Burlington House influence on painting. Covent Garden Opera has never been more influential than just before the Four Years War, and that same war found a group of English poets and novelists at their best. As a baronet's son, Beecham was able to command a following among the upper middle classes, but he soon found that the limitations of the New Symphony Orchestra restricted his repertoire and prevented the accomplishment of his artistic purpose. The orchestra was induced to become larger, and it was at this point that it became a limited liability company; but the players kept alive the deputy system. This upset Beecham's artistic designs again, and he took the step in 1909 of forming his own Beecham Orchestra.

On its foundation the Beecham Orchestra exposed a weakness in the organization of earlier London orchestras. One result of insistence on good sight-readers was a preference for players of long experience. Young players leaving the academies found difficulty in obtaining engagements. Here the deputy system was of some use, since it was the only way by which these players could prove their worth before a conductor. Orchestral players, however, were wary, and as a matter of policy avoided sending deputies who played better than they themselves. The plight of the most capable young players was therefore the worst. Beecham staked his future on youth and enthusiasm instead of on guile and experience: the average age of the players in the Beecham Orchestra was under twenty-five. Albert Sammons, who soon became his leader, Beecham found playing in the Waldorf Hotel. Another discovery was Lionel Tertis, the first viola virtuoso. Among these players Beecham found an absence of strongly-held views on the essentials of orchestral playing, and he was able to train them quickly in a distinctive style that we have since come to know as Beecham's own. There was a freshness and sparkle about their playing that won the admiration of audiences everywhere. Soon the Beecham Orchestra began to fit into the niche its conductor

THE TWENTIETH CENTURY ORCHESTRAS

intended it should occupy in London musical life. A season of grand opera was announced at Covent Garden Theatre where Strauss's *Electra*, Delius's *A Village Romeo and Juliet*, Sullivan's *Ivanhoe*, Ethel Smyth's *The Wreckers*, Bizet's *Carmen*, Humperdinck's *Hansel und Gretel* and Debussy's *L'Enfant Prodigue* were produced. *Electra*, like anything new by Strauss in those days, was a popular success in the best musical circles, for reasons which Beecham understood well enough, for he is reported to have said that the people who ran to see *Electra* would have run with even more zeal to see an elephant standing on one foot at the top of the Nelson Column. Ethel Smyth, instead of throwing in her weight to help Beecham, made cutting remarks about opera in England and stated that she had recently refused two offers for *The Wreckers* to be produced in England.

In the following year Beecham gave a season of light opera at His Majesty's Theatre, had an amusing skirmish with the Lord Chamberlain's office about the decencies in *Salome* and *Der Rosenkavalier*, and prepared for the most important event in the history of artistic achievement in England during the twentieth century. This was the introduction to London of the Diaghileff Russian Ballet.

The Russian Ballet brought a new conception of that art — or rather combination of arts — by way of Paris to London. The old conventions of ballet, under which music and scenery had been subordinated to the dancers, and the male dancers had declined into a role that supported the solo female dancers, were thrust aside in favour of an entertainment in which music, choreography, and painting all combined in an original dream-fantasy of Russian fairy-tale. It was with the Beecham Orchestra, also, that Diaghileff introduced his Russian Ballet to Germany in 1912 — outflanking that country's anti-slav cultural campaign with a movement from the west. There Beecham made the favourable impression in Berlin that has already been mentioned. The Diaghileff Ballet made London its headquarters, and a period of opera combined with Russian Ballet followed. *Boris Godounov*, *Khovanstchina*, and *Ivan the*

THE NATIONALIST PERIOD

Terrible, with Chaliapine, are landmarks in operatic history, while the first performance of Stravinsky's *Petrouchka* was only one of many triumphs for the ballet. There had been many previous performances of the Russian nationalist compositions in the concert hall, for which credit must go mainly to Sir Henry Wood; his first wife, a Russian singer of excellent taste, brought the works of such composers as Glinka, Borodine, Mussorgsky, and Rimski-Korsakov to his notice, and enabled him to extend the public's knowledge of Russian music which had up to that time centred round Tschaikovski and such other Russians as had been acceptable to the Germans. Beecham was the first to stage the operas of these men in London, and perform their work in their entirety. On the outbreak of the Four Years War, London was no longer in an inferior position for the production of opera — whatever Ethel Smyth might say — for 1914 saw a remarkable season of opera given by the Beechams, father and son. Sir Joseph Beecham undertook the business management of the enterprise while his son took charge of the artistic arrangements. Even the war did not stop Beecham's musical activities, though it laid its deadly hand on most. In addition to opera in London, Beecham gave valuable aid to the Hallé Orchestra, and weaned Manchester from the diet of German music to which it had grown accustomed under Richter. These activities upheld British music in her darkest hour, but the financial burden was too much for the Beecham resources, great though they were. The death of Sir Joseph came as a last severe blow, and Sir Thomas (who had, of course, succeeded to the title) was forced to spend the three years from 1920 to 1923 in a struggle for financial recuperation that forbade music. By the time he was able to resume his musical life the scientist had stepped in, and the relationship between audiences and performers was on the verge of divorce.

REFORMATION

AMONG the more valiant of Sir Thomas Beecham's efforts during the Four Years War was an attempt to establish orchestral concerts on a secure basis in Birmingham. The result was a serious financial loss that can be attributed mainly to the commandeering of the Town Hall by the War Office; but this particular trial might have been borne with more fortitude had less cultural opposition been encountered. As a choral centre, Birmingham enjoyed a considerable reputation, yet even Bantock — who among other claims to fame had been instrumental in introducing Sibelius first to British audiences, and had brought the composer to Birmingham in 1912 to conduct there the first performance of his *Fourth Symphony* — even Bantock had not been able to establish a first-class orchestra permanently in that city. J. G. Halford had worked hard with an orchestra partly amateur and partly professional, but had finally retired from Birmingham a disappointed man, crushed between the upper millstone of expert criticism and the nether millstone of choral society prejudice. These societies represented a majority opinion, and in many provincial towns abused their power. Higher aesthetic standards necessarily attract fewer adherents, and, moreover, choral societies were numerically larger than orchestras; they utilized large forces of unpaid labour and were consequently more easily maintained than an orchestra. The choristers paid the pipers, and to them the orchestra was the 'accompaniment'; even so famous a choral conductor as Sir Henry Coward of Sheffield had performed Handel's *Messiah* with the orchestral codas omitted, so that the orchestra came to a stop on the chord on which the chorus or soloist finished. The most common insult, however, was the refusal of many choral societies to associate with orchestras from their own localities, preferring scratch orchestras or imported orchestras. The usual psychological reaction

THE NATIONALIST PERIOD

took place—just as British musicians had stiffened their ranks against foreigners who condescended to take money from *das Land ohne musik*, so orchestral societies stiffened their ranks against the semi-skilled choral hordes. But here the reaction was ineffective because co-operation was essential for success. Accusations of ignorance made against choral opinion hit back like boomerangs, for choirs were a source of great local pride, while local orchestras were generally far inferior in skill to local brass bands. The reason was educational: both the brass bands and the choirs had the benefit of simplified notations, but the orchestral player had to learn the hard way, and consequently made slower progress, although his knowledge was more complete. So it happened that outside London only Manchester, Bournemouth, and Scotland had first-rate orchestras; great centres of population like Birmingham, Leeds, Sheffield, and Liverpool had only struggling orchestras, but were renowned for their choral activities.

The problems arising out of civic pride in choirs and brass bands with their corresponding contempt for local orchestras can only be solved by local action; visits of famous orchestras from elsewhere tend only to reinforce this contempt. Birmingham, however, attacked the local orchestral problem resolutely at the end of the Four Years War while a craze for social reconstruction was in evidence. Their resolution was aided by certain caustic remarks from Sir Thomas Beecham, which brought their usual patriotic reaction, reinforced by an inward conviction that they were true. Foremost in the fray was Appleby Matthews, a Birmingham musician trained at the Midland Institute, who was in 1918 attracting attention as a solo pianist, choir conductor, and conductor of the Birmingham Police Band. The latter, a wind band, was brought into being by a musical chief constable; its members, mostly drawn from the Royal Military School of Music at Kneller Hall, were on the staff of the Birmingham police. The band used low pitch, although that did not come into use officially in military bands until 1927. Matthews chose a suitable group of wind soloists

REFORMATION

from his police band, added a string section, and entered upon a series of Sunday concerts at his own risk in 1918. Sunday had to be chosen because Matthews drew many of his strings from the professional players employed in theatres and cinemas during the week, and Sunday was their only free day. Had Birmingham been one of those cities where a sabbatarian majority can veto Sunday entertainment, Matthews' venture could not have taken place, but Birmingham is a mystery to outsiders: politically conservative in those days, it nevertheless established a Municipal Bank; its morals gave offence to a famous ballet dancer, because a by-law demanding that every dancer should wear tights in Birmingham was rigidly enforced; yet the City Art Gallery was opened on Sundays in order that young people, who might otherwise have been obliged to walk the streets, might perambulate in comfort and possibly experience the benefits of aesthetic sublimation; and the Corporation encouraged Sunday concerts. More; in 1920 the City Council, led by Mr. Neville Chamberlain, granted £1250 a year from the rates for the purpose of maintaining a Municipal Orchestra, and this grant supplemented by numerous private donations enabled Appleby Matthews to reorganize his orchestra on a permanent basis.

Preference was given to those players who had already assisted the venture by giving up their Sundays to rehearsals and concerts with little material reward, for the proceeds of Sunday concerts are earmarked for charities. Credit must be given to those players who, although condemned to work long hours without pause in a darkened cinema, eyes strained by the glare of an illuminated music-desk set in the midst of a blackness relieved only by the light from a flickering screen, and ears strained by the task of making the smallest of orchestras produce a tone that would penetrate to the far reaches of a large auditorium, yet gave their free time to Matthews for the establishment of his orchestra. Unfortunately, when the time came for them to be offered the more congenial employment of symphonic work, many of these players were unable to

THE NATIONALIST PERIOD

accept it for financial reasons. Hackwork may destroy the soul but it fills the belly, and it cannot be too often impressed on those who judge conditions from the starch and ceremony of the concert hall, that the performance of good music has always involved some sacrifice from the rank and file. Star artists only get an unfair share of the concert-goers' money.

The Birmingham City Orchestra started with a programme of six symphony concerts a year (later increased to eight), six popular Saturday concerts, and twenty-four Sunday concerts. In addition, the future of music in Birmingham was aided by the provision of six Saturday afternoon concerts each year for school children. Considerable losses were sustained, so that the orchestra entered on its 1924-25 season with an adverse balance of £3000. This was the position when Dr. Adrian C. Boult took control. Financial stability he could not bring, nor has the city demanded it — indeed, municipal support of orchestral music in Birmingham has been generous, no less than £14,500 in 1944 — but Sir Adrian Boult established orchestral music in the city on a reputable artistic basis from which it has never looked back. Under Dr. Boult (as he then was) a young violinist named Paul Beard came to the fore as an orchestral leader, in which capacity he was later to serve Sir Adrian in the wider operations of the B.B.C. Symphony Orchestra. To both these men British music owes a great deal.

Like Sir Thomas Beecham, Sir Adrian Boult hails from Liverpool. There he conducted twelve concerts in the season 1914-15 and a further six in 1915-16, using a small orchestra of Mozartian symphonic dimensions with a foundation of fifteen strings. Liverpool, however, has shown itself repeatedly unable to retain good conductors, and it was Boult's appearances with the orchestra of the Royal Philharmonic Society in 1918 that marked him out as a conductor of exceptional merit, particularly his first public performance of Holst's suite, *The Planets*. Not all Holst's suite was played on that occasion — *Venus* and *Neptune* were omitted, but the remaining movements — *Mars*,

REFORMATION

Mercury, Jupiter, Saturn, and *Uranus* — revealed a variety and a mastery of orchestral resource in both composer and conductor that would have drawn attention to this remarkable manifestation of British musical virility even if the inflated patriotism of the war years had not favoured a bias in that direction. Patriotism further revealed itself in orchestral life immediately after the war, when the British Symphony Orchestra came into being. This was organized by players released from service in the armed forces, and was conducted by Boult. Such an orchestra was, however, but an addition to the number of similar orchestras of longer standing, such as the London Symphony Orchestra and the New Symphony Orchestra, now appearing under the name of the Royal Albert Hall Orchestra and conducted generally by Landon Ronald or Sir Thomas Beecham (after 1923). All these orchestras, together with those appearing under the names of the Royal Opera Orchestra and the Royal Philharmonic Orchestra, held in common a good percentage of their personnel — the operation of the deputy system — and were coming to be regarded by their members as avenues to employment rather than as separate orchestras having each a distinct style. They offered little scope for a conductor to develop a fine style, since that can be done only by regular rehearsal with the same personnel over a term of years. Such an opportunity Boult seized when he became conductor of the Birmingham City Orchestra in 1924, and he was able during the six years he had it under his control to mould this orchestra into one worthy to serve not only Birmingham but also a surrounding tract of Midland towns which the Hallé and the London orchestras found equally inconvenient.

One orchestra that started up in 1924 had a quite distinct personnel: this was the British Women's Symphony Orchestra. During the war many excellent women musicians had replaced men in our orchestras, but with the return of men from the services these women were displaced. There arose in consequence a protest among these women against such sex-distinc-

THE NATIONALIST PERIOD

tion, which they countered by forming their own orchestra on the now familiar London Symphony Orchestra co-operative basis. They were conducted during their first year by Miss Gwynne Kimpton, and in the following year, 1925, by a young conductor of great promise named Dr. Malcolm Sargent. He was a man of fine intellect, at that time doing valuable work as a conductor of symphony and opera, and on the teaching staff of the Royal College of Music.

It was a time of reconstruction, not of distinction, in orchestral circles. Progress was best made in musical education and in composition in the decade following the Four Years War. Composers seemed determined to cut adrift from everything to which they had been tied in the past, knowing that they would be storm-tossed, but believing they would nevertheless be free. Not a few foundered, some made port either in this country or the U.S.A.; the unfortunates were certain minor composers of the Edwardian era who found themselves actually becalmed by 1920. Reconstruction of musical education was much better planned, however, and achieved a bloodless revolution through the efforts of an army of local educational authorities subsidized without their awareness by the advertising departments of important vested interests. Not since the eighteen-forties had such a concentration of attention on musical education for the masses taken place.

Nineteenth-century advances in musical education had been inspired by pianoforte manufacturers and composers, like Clementi, who encouraged the rich to have their daughters instructed in keyboard technique, and humanitarian reformers and sheet-music dealers who encouraged the poor to sing. Twentieth-century educational policy was revolutionary because it transferred attention from skill as performers to skill as listeners. As early as 1910 Stewart Macpherson had been advocating education in musical appreciation, but the most effective impetus came after the Four Years War from a series of small books issued by the Oxford University Press, notably those by Dr. Percy Scholes and *A Musical Pilgrim's Progress* by

REFORMATION

an amateur named J. D. M. Rorke. The appearance of these publications coincided with an intensive campaign of advertisement by manufacturers of gramophones, player-pianos, and, later, radio receivers. Soon there was general agreement among manufacturers and musical educationalists of the highest standing that the ability to listen to good music intelligently was of more importance than to play it badly; this view found its highest expression in a summer training course for music teachers at Oxford, organized jointly by the British Music Society and the Federation of British Music Industries. Musical form and history took pride of place in published books on appreciation, but the repetitive character of mechanical musical instruments directed attention to the business of score-reading. Interest in orchestral colouring, which had suffered in the old days when familiarity with orchestral scores in the home could only be got through transcriptions of orchestral works for piano duet, now quickened, and the orchestra gained in interest even in towns where opportunities of hearing it at its best were rare.

Thus did the gramophone justify itself in spite of its sandpaper tone and the distorted balance of the old acoustical recordings. It was found that the serious music-lover had an amazing aural adaptability, being capable of ignoring the scratch of the needle to an extent that is almost unbelievable to-day; so much so that the player-piano, which did not distort pianoforte tone, lost favour because of its robot-like interpretations, while the gramophone actually increased in favour with pianoforte music-lovers. To a very great extent this favour was due to the superior advertising skill of the gramophone-record manufacturers, who induced the world's most famous musical celebrities to lend their names to eulogies of the new recordings, which were naturally issued on a royalty basis. When broadcasting by wireless telephony began with the British Broadcasting Company in 1922, these celebrities were at first less willing to accept the new medium, but the publicity value of the novelty soon convinced them that it was a boon to their art.

THE NATIONALIST PERIOD

Within a few years the musical profession was enslaved by alternating current as effectively as the craftsmen of the eighteenth century had been enslaved by steam. Even those whose consciences had always put their art before all other considerations became involved. How the microphone came into the orbit of Sir Thomas Beecham is a case in point; he had promised to conduct and speak at a concert in Manchester, the proceeds of which were to go to the Hallé Pensions Fund. When he got there he found that he had either to disappoint the players or submit to the microphone. In reporting the speech the *Manchester Guardian* omitted some of his remarks on the nature of monopolies and on the nature of radio reception, but sufficient remains for an unvarnished picture of the orchestral scene to be presented. Sir Thomas said that

> 'The orchestra was Manchester's leading musical institution, and in his own view the city ought to be proud of it. But everything in this country concerned with large and serious musical institutions was in a condition of danger and stress ... Every month and every year took from us a large number of our best players.' (They went to the U.S.A.) 'Those American orchestras were founded and endowed on a generous and even a lavish scale that was quite beyond the possibilities of this country ... But he thought we could do something more for our orchestras than we did at present. In years gone by an orchestra, even if it gave its players a limited number of engagements, was an important thing to them, but since the advent of cinemas and music-halls, café orchestras and other "diabolical distractions" it was no catch to play in a symphony orchestra. You may be under the impression that by attending these concerts you are doing something that reflects credit on you', he said, 'that in some way you are patronizing good orchestral music. The boot is on the other foot. It is such a miserable business playing good music that orchestral players are patronizing you and doing you a great service. Orchestral players of first-rate quality frequently play at good concerts at considerable sacrifice.'

REFORMATION

Sir Thomas then gave some facts about the Hallé Pensions Fund, pointing out that the net sum payable to a pensioner on retirement was £25 a year; the report continued:

> 'I have come here at enormous inconvenience', continued Sir Thomas, amid laughter and applause; 'I have crossed the channel; I have been ill; I have caught a cold; I have postponed three concerts on the continent. If I do that, I think you might do something equivalent.
> 'Another thing: sometime ago I vowed eternal warfare on the wireless. To me it is one of those modern inventions to which I shall never accommodate myself. But I made a solemn pledge to myself to speak into that instrument tonight, and so I have violated one of my sternest principles. You cannot do too much in these days to keep alive the serious musical organizations that you have. I see the orchestra, every month and every year, getting a little feebler and more rickety in its foundations and support.'

It was true. Only too often the music-lover preferred the loudspeaker's interpretation in the comfort of his home to the real thing in a draughty auditorium, and audiences were accordingly capricious. Worse: chamber music suffered. Wealthy hostesses balanced the economy of chamber music on the wireless against the expense of having it played to them in their homes. In the Hallé Orchestra at that time were three players — Arthur Catterall, the leader; John S. Bridge, the principal second violin; and Frank Park, the principal viola — who, with Johann C. Hock, a Birmingham 'cellist, formed a string quartet of some fame. The Catterall Quartet supplied the demands of chamber music enthusiasts in the North much as the Hallé Orchestra supplied the demand for orchestral music. Indeed, the Quartet went further, for it served some of the territory rightly belonging to the Scottish Orchestra. Each year the Catterall Quartet could rely on about forty engagements from the owners of great houses in the North of England and the Lowlands of Scotland — not casual engagements from

THE NATIONALIST PERIOD

strangers, but regular yearly bookings by the same people; people whom the members of the Quartet had come to regard almost as personal friends. Yet within five years of the coming of the loudspeaker these patrons — the cream of musical taste in the North — had dwindled until the continuance of the Catterall Quartet could no longer be justified on financial grounds. Radio reception favoured small combinations of instruments, but it was something of a surprise to find that loyal patrons of chamber music preferred the spark-splashed product to the real one. Their support of so completely perfect an ensemble as the string quartet had led to a belief that they were gifted with exceptionally sensitive hearing; even so, they could subordinate their taste to their pockets. In London the truth was brought home to us most forcibly in 1927-28 when the Berlin Philharmonic Orchestra visited this country and put our orchestras to shame.

Yet the interest in orchestral music continued to grow. The British Broadcasting Company had quite early in its career found that an orchestra was more suited to its general needs than any other instrumental combination, and the orchestra therefore formed a background to home life throughout the country. Broadcasting, however, was not the only means of propaganda favouring the orchestra, for a concert impresario named Lionel Powell organized an extensive scheme of International Celebrity Subscription Concerts in which the London Symphony Orchestra and the Royal Albert Hall Orchestra took part, appearing in most large towns under the batons of Sir Landon Ronald, Sir Thomas Beecham, Albert Coates, and, of special interest, the London Symphony Orchestra appeared at these concerts in 1930 under Mengelburg. It was possible for provincial audiences to compare styles in orchestral playing in a way that had not been generally possible previously outside London.

There was a special reason for Mengelberg's appearance with the London Symphony Orchestra. So impressed were the players by the superiority of the Berlin Philharmonic Orchestra

REFORMATION

in London in 1927 that they engaged Mengelberg to improve their style as he had done that of his famous Concertgebouw Orchestra of Amsterdam. 1927-28 can, in fact, be regarded as a time of reconstruction in British musical affairs, for in that year the British Broadcasting Company gave way to the British Broadcasting Corporation, a public corporation operating under Royal Charter with a monopoly of broadcasting in the United Kingdom, the Channel Isles, and the Isle of Man. The new Corporation was controlled by a Board of Governors appointed by the King in Council, and answerable to Parliament through the Postmaster-General. The economic basis of music-making was completely transformed, but, as in the case of the London Symphony Orchestra, the practical effects did not begin to show themselves until 1930.

In that year Adrian Boult was appointed Musical Director of the B.B.C. and founded the great orchestra that bears its name.

Just as the Germans had learned to keep a good display in their western shop-window, now it had become necessary for the British to do the same. People in any part of the world could form an opinion of British taste by the turn of a button, and the impression they got during the 'twenties was generally unfavourable. The justification for a monopoly in broadcasting was that a unified British policy might have at some time to be presented to the rest of the world; the loudspeaker could not help but be an instrument of propaganda; it became necessary therefore to organize a British orchestra that would worthily present the British to the rest of the musical world. Advantageous terms were offered to orchestral players by the B.B.C. — terms which no other orchestra could offer — and by this means an orchestra of first-rate players was got together.

Boult inaugurated a comprehensive scheme for a full orchestra of 119 players capable of subdivision into four sections called B, C, D, and E. The ingenuity of his plan will be further understood on perusal of the following table:

THE NATIONALIST PERIOD
B.B.C. SYMPHONY ORCHESTRA

	A	B	C	D	E
First Violins	20	14	6	12	8
Second Violins	16	12	4	10	6
Violas	14	10	4	8	6
Violoncellos	12	8	4	8	4
Double Basses	10	8	2	6	4
Flutes	5	3	2	3	2
Oboes	5	3	2	3	2
Clarinets	5	3	2	3	2
Bassoons	5	3	2	3	2
Horns	8	4	4	4	4
Trumpets	5	3	2	3	2
Trombones	6	3	3	3	3
Tuba	1	1	—	1	—
Tympani	2	1	1	1	1
Percussion	3	3	—	2	1
Harps	2	1	1	1	1
	119	80	39	71	48

B + C = 119   D + E = 119

From these it will be seen that when the full B.B.C. Symphony Orchestra is not in use, an orchestra of classical symphonic dimensions can be employed at the same time as an orchestra sufficiently large for the majority of later compositions, and that there are two alternative ways in which this can be done. Such a scheme was far beyond the means of any other orchestral organization in this country, but even so it was not overelaborate for broadcasting needs; indeed, the B.B.C. Symphony Orchestra was a hard-worked body, certainly not exorbitant for the duties it was called upon to perform.

With this orchestra under Boult's admirable direction a high standard of performance was rapidly achieved, and the prestige of musical Britain no longer need be feared. In the process of forming the B.B.C. Symphony Orchestra, however, considerable dislocation was sustained by every other symphony

REFORMATION

orchestra in the country, for Boult chose only the best players, most of whom belonged to other orchestras. So great was the loss sustained by the Hallé Orchestra — which involved their leader, Catterall, their first oboe, Whittaker, and their first bassoon, Camden — that Sir Hamilton Harty, who had conducted the Hallé since 1920 and brought it to a pitch of proficiency worthy of that orchestra's best traditions, resigned his post. To the credit of the B.B.C. it must be said that they sympathized with the Hallé Society in its misfortune, and strove to assist its recovery by permitting players from the B.B.C. Northern Orchestra to appear with the Hallé Orchestra, and broadcast the Hallé concerts, but the less said about the B.B.C. Regional Orchestras the better, for they perpetuate a pre-Boult policy and remain unworthy occupants of our British shop-window. The recovery of the Hallé Orchestra is due to the superlative skill of its present conductor, John Barbirolli.

In addition to the concern of the B.B.C. for a better standard of orchestral playing must be mentioned that of Sir Thomas Beecham. Indeed, before Boult's association with the B.B.C. Sir Thomas had discussed with Sir John Reith a plan for an orchestra that would serve both for broadcasting and for concert performances, but agreement could not be reached on the part to be played by outside concert organizations. Sir Thomas then turned again to the London Symphony Orchestra, which he had been conducting intermittently since his return from the U.S.A. in 1928, and endeavoured to reach agreement with them on a policy of reconstruction, but his scheme failed to find acceptance because it would have involved changes of personnel to which the players might not agree. Beecham therefore found himself obliged to form a new orchestra to his own liking; this he did in 1932, drawing his players, as in 1910, mainly from the ranks of young musicians, but, fortunately for the later history of the orchestra, obtaining also the services of some London Symphony Orchestra members who had had experience in co-operative orchestral management. Sir Thomas entered into an arrangement also

THE NATIONALIST PERIOD

with the Royal Philharmonic Society by which his new orchestra should play at their concerts, and he called his new combination the London Philharmonic Orchestra.

This use of the title 'Philharmonic' was essential for Beecham's purpose because he intended that it should invite comparison with other orchestras of that name abroad. In spite of the ravages of the B.B.C. Sir Thomas had got together a very promising body of men, which included Paul Beard as leader and Leon Goossens as first oboe; the orchestra was registered as a limited liability company with a board of directors well known in the City as able business men, and in Society as staunch champions of music.[1] After twelve rehearsals the London Philharmonic Orchestra appeared at Queen's Hall on October 7th, 1932, with a programme of works by Berlioz, Mozart, Delius, and Richard Strauss. Of it Ernest Newman wrote that 'nothing so electrifying has been heard in the London concert hall for years. The tone was magnificent, the precision perfect, the reading a miracle of fire and beauty'. And from this beginning the London Philharmonic Orchestra went on to justify its conductor's hopes. Orchestral music came to the fore in London society, and the opera again showed decided improvement at Covent Garden. A visit to Brussels in 1935, a tour of Germany in 1936, and a visit to Paris in 1937 left favourable impressions of British music in those countries, but certain suspicions of political colouring in German musical opinions made Sir Thomas wary of accepting an invitation to tour Italy in 1938, and the project was dropped. The B.B.C. Symphony Orchestra visited Vienna, playing under Sir Adrian Boult, and there were visits to Britain of the Berlin, Vienna, and Prague Philharmonic Orchestras. Whatever their foreign policies might suggest, European countries showed themselves almost pathetically anxious to prove through music their nominal sympathy with civilization.

Foreign philharmonics were of course subsidized from State funds. The London Philharmonic had no subsidy and no

[1] Two of these resigned in 1936.

REFORMATION

reserves. This orchestra played so magnificently because Sir Thomas Beecham placed artistic supremacy before everything else — even financial stability. It was not thought that the orchestra would long withstand the effects of any economic crisis that might come, for since 1935 the players had been paid a regular salary instead of, as previously, on a piecework basis. War came in 1939, and with it the financial collapse of the Philharmonic. Sir Thomas Beecham left for the U.S.A. and the members of the Philharmonic Orchestra reorganized their source of livelihood under the title of Musical Culture Limited. In so doing they were able to use the services of some of their members with experience of committee work with the London Symphony Orchestra, and they found among their violists Thomas Russell, who proved to be a secretary with imagination, drive, and a zeal for hard work. He has described the struggle of the London Philharmonic Orchestra through the war years in *Philharmonic Decade*; it is a grim record of desertion by former friends, a cat-and-mouse ordeal with the Ministry of Information, of friendly welcomes in bomb-shattered towns, and encouragement from two men who approached a democratic people from opposite directions — Mr. J. B. Priestley, the novelist, and Mr. Jack Hylton, the dance-band leader and theatrical manager.

It may be asked what justification had Sir Thomas Beecham in 1932 for starting again a chain of circumstances that might easily lead to the same triumphs and the same disaster as before, when science and popular demand had combined to solve the old problem of cost, and the benevolent monopoly that resulted had organized a first-class British orchestra? The answer is that there can be no single solution to any problem of taste. Sir Thomas Beecham's ideals of orchestral performance differ widely from Sir Adrian Boult's, but there is no question of right and wrong about these ideals: it is the nature of men of fine taste to be individualists in their art-work, and society is the richer for it. Indeed it may be said that without such individuality of taste there is no true art. The marked difference

THE NATIONALIST PERIOD

in style of the London orchestras has been summed up thus by Frank Howes:

> The connoisseur of to-day may be able to detect, even behind the dazzling façade of the conductor's personality, something of the corporate life of the orchestras to which he listens either in the hall or on the gramophone record. Actual quality of tone, especially in the wind departments, depends largely on individual instrumentalists, yet when the balance is made up, orchestras offer to the ear of the listener their own peculiar sonority. It has for instance been noticeable during the last years of the decade of the nineteen-thirties that the three London orchestras have each a distinctive quality, as well as a general attitude to music, which affects their style of playing. It may be metaphorically described in terms of drapery; the L.S.O. sells a cotton textile, the L.P.O. a silk, and the B.B.C. a velvet. Their specific virtues are: of the L.S.O. adaptability and resource, of the L.P.O. a lyric, singing style, derived from its conductor's musical mentality, and great finesse; of the B.B.C. massive power and executive brilliance both corporately and individually. Their compensating defects are: of the L.S.O. a certain colourlessness that sometimes degenerates into laxity (of attack, ensemble and phrasing); of the L.P.O. an acuity of tone and attack that sometimes hurts the ear with its assaults — the strings might be whipcord lashes; of the B.B.C. an inflation that sometimes sounds heavy, sometimes bloated.[1]

Criticism can be extended to include orchestras in other parts of the country, for upon their individuality depends their survival. There is considerable local pride in provincial districts that have their own orchestras, and they will no more submit to the idea that a centralized interpretation of music is authentic than they will admit that B.B.C. English is preferable to their own. The lesson of the Six Years War is that the British react even more strongly to circumstances that tend to deprive them of the type of music they want, for during that

[1] Frank Howes, *Full Orchestra*.

REFORMATION

period the popularity of orchestral music increased all over the country to an extent certainly not foreseen. The facts are plain enough, but an explanation is by no means simple.

At the end of the Four Years War Mr. J. Mewburn Levien, who was then secretary of the Royal Philharmonic Society, estimated that in London 3000 people were permanently interested in orchestral concerts, and that of these 2500 were members of his society.[1] Queen's Hall, with a seating capacity of 2492, had always been adequate for these concerts and also for Sir Henry Wood's Proms. Indeed, in 1927 Messrs. Chappell & Co. found themselves unable to continue the Proms. and the B.B.C. stepped into the breach to prevent so admirable an institution from coming to an end. Private enterprise had failed to keep going a popular effort in the face of competition from the new machine. The end of Queen's Hall came on the night of May 19th, 1941, with a direct hit by a German bomb. The next season of Promenade Concerts had to be held in the Royal Albert Hall, which has a seating capacity of 8000, and before the end of the Six Years War it had become obvious that the Proms. audience would fill an even larger hall if there were one. This happened, moreover, in times of the greatest inconvenience, with inadequate transport, dark streets, and in constant danger of bombs. There was no diminution of attendances when Sir Henry Wood died in 1945, and a great improvement in the standard of the playing when Sir Adrian Boult divided the season between two orchestras, thus relieving the strain on the players. The youthful character of the audiences was more marked than ever, with a great number of people in uniform. What was true of London was true of other parts of the country, except that in many places the concert halls were not big enough to accommodate all who wished to enter.

An explanatory theory must start from the musical appreciation movement at the end of the Four Years War, but exposes a fallacy. Musical appreciation was founded on the knowledge

[1] Mr. Levien tells me that he got this estimate from Robert Newman.

THE NATIONALIST PERIOD

that good music retains its appeal whereas superficial music loses its appeal on familiarity; from which it was argued that if sufficient opportunities for often hearing the best music were made available, people would come to prefer it, and would reject that which on repetition palls. They ignored the question of taste. There was soon no dearth of opportunity to hear all kinds of music through the ether, and the appreciationists' bluff was called, for throughout the country choral societies got into financial difficulties, the Proms. lost money in London, while superficial music rolled off the presses, spread like a forest fire and was discarded in favour of another 'hit' in no more time than the currency of a popular film. The machine not only provided opportunities for the repetition of good music, but for rapid changes of bad music, so all were equally satisfied. Serious music, however, is more introspective than superficial music, so a greater proportion of serious music-lovers stayed at home and got their music through the machine, and it was impossible adequately to estimate the amount of favour with which it was received. Then came the war, and the serious music-lover found himself caught up in the national effort like everyone else; thousands were thrust into a new communal way of life which had little regard for personal tastes, and where the superficial minds were in the majority. The canteen or hostel radio blared out all day, and only the news was acceptable to everybody; in music it obeyed the will of the majority, which was anathema to the serious music-lover. He accepted the situation but reserved the right to disagree with the majority, and took to concert-going, at whatever inconvenience, whenever the opportunity presented itself. With a great deal of hard work by travelling orchestras like the London Philharmonic Orchestra he was satisfied, but let it not be thought that the proportion of serious music-lovers to superficial ones was greater than it was in, say, 1910; the minority had become unified in self-defence, that was all.

This unification of minorities in the face of a threat to their tastes is a prominent feature of British life, and indeed it is one

REFORMATION

of the virtues of democracy that minorities have a right to self-expression. So natural has it become that we do it without thinking. It needs a spur to start a minority into unanimity, however, and the best spur is a sense of grievance, real or imagined. That is what happened in Liverpool. As a result of war organization, Giant Monopoly shuffled, and twenty-five players, members of the B.B.C. Northern Orchestra, who had previously been engaged by the Liverpool Philharmonic Society for their concerts, were forbidden to accept such engagements in future. Yet they were permitted to accept engagements with the Hallé Orchestra. The pill was hard to swallow because there has always been a sense of civic rivalry between Manchester and Liverpool. Liverpool opinion hardened, and turned towards its own resources. There had been a professional orchestra in the district since 1930 organized and conducted by a local violinist named Louis Cohen, called the Merseyside Orchestra. Its activities in Liverpool were mainly Sunday concerts for the people in St. George's Hall, chamber concerts in the Walker Art Gallery, and assistance at choral concerts over a fairly wide area. The Liverpool Corporation had encouraged this orchestra by allowing them the use of St. George's Hall at a low rent. St. George's Hall, however, was not so good as the Philharmonic Hall, which belonged to the Philharmonic Society. The war, which deprived the Liverpool Philharmonic Society of the right to engage certain players, deprived the Merseyside Orchestra of St. George's Hall, which was wanted for purposes of national defence. What more reasonable than that they should combine forces, the exclusive dignified Philharmonic Society and the professional orchestra of forty-two players that gave music to the people? They did so, the orchestra was enlarged to contain sixty, the Corporation entered into the scheme with an agreement to purchase the Philharmonic Hall, and to grant the free use of the hall and a subsidy of £4000 a year. At first the popular concerts continued to be conducted by Louis Cohen and the week-day concerts by guest conductors, until the permanent appointment of Dr. Malcolm

THE NATIONALIST PERIOD

Sargent. Far from being neglected, the end of the Six Years War found the Liverpool Philharmonic Orchestra an overworked organization. It was not so overworked, however, as the London Philharmonic Orchestra, which drew no subsidies except a small grant of about four per cent of their costs from the wartime Council for the Encouragement of the Arts, and had to tour the country 'giving concerts all day and every day', as Thomas Russell says, 'with endless travelling, little time for rehearsal and no time for relaxation'. With the increased support that came towards the end of the war some improvement became possible, but at best financial equanimity demanded six concerts a week and three rehearsals. Under these conditions the players' lives were an artistic nightmare. Some lowering of the standard Beecham had previously attained had to be tolerated, and with so little time for rehearsals first performances of new works were not sought.

In spite of a grant of £2500 from Manchester Corporation (mainly for educational concerts) the Hallé Orchestra found itself in a position little better than its founder had experienced — players had to be overworked in order that the guarantors of the Hallé Society should not be called upon. And this was the state of orchestral music in England when audiences were more plentiful than ever before! Quite well directors of orchestras knew that an economic slump would reverse the position and bring disaster.

The Scottish Orchestra continued to justify its name, and found that their countrymen appreciated this, for besides the large grant the orchestra received from the Corporation of Glasgow, many of the smaller Scottish towns made contributions according to their means, and regarded the orchestra as a national Scottish responsibility. The same unanimity of patriotism could not be expected from the English counties (though they were doing more in direct assistance from local taxation than London) and Sir Dan Godfrey, whose advice on municipal management would have been invaluable, died in 1939.

REFORMATION

A spontaneous private mark of enthusiasm which had first appeared in 1933, however, came into prominence. This was the Proms. Circle, a group of Queen's Hall promenaders who met together in the close season to keep in mind the pleasures they enjoyed so much. This they did with monthly recitals of gramophone records and occasional lectures. Sir Henry Wood found in them much gratification: a 'guardianship', he called them. 'In you, I see a guardianship of inestimable value, not only associated with the Promenade Concerts, but as enthusiastic connoisseurs of all that is best in music, and as curators of the fine tradition associated with the great orchestral repertoire, which I have endeavoured to uphold and encourage during my long life of music.' Ten years later came the Philharmonic Arts Club with meetings twice a week and, in London, a very healthy interest in contemporary music. This club cannot be said to have arisen as spontaneously as the Proms. Circle, for it was started by the London Philharmonic Orchestra as a means of breaking down the reserve that separated the orchestra from its audiences. It rapidly became a lively organization and set an example to many provincial districts, where local clubs of a similar nature were formed. The association of such clubs up and down the country, with an interchange of lecturers and an official magazine would be in the normal course of the democratic tradition, and might reasonably be expected to play a great part in the history of the orchestra in England, especially as its counterpart in orchestral administration came into being in 1943 with the formation of the National Association of Symphony Orchestras, comprising the Hallé Orchestra, the London Philharmonic Orchestra, the Liverpool Philharmonic Orchestra, the London Symphony Orchestra, and the Scottish Orchestra. The difficulties of organizing the Association in its early stages were overcome with the assistance of the Musicians' Union.

From the beginning it was laid down that the term 'symphony orchestra' meant an organization of not less than sixty players, and that no smaller orchestra could be admitted to the Association. This was done not with any intention of making the Association exclusive but in order to protect the public against cheese-paring organizations whose claim to recognition lay in the high-sounding title of 'symphony orchestra', and not against such excellent orchestras as the Boyd Neel String Orchestra and the Jacques String Orchestra. It may be mentioned that the orchestras that founded the Association are all considerably larger than the minimum laid down in the rule.

It can be seen that there is emerging a general policy in British musical life, with links between performers and audiences welded on a democratic anvil, making for progress from an acknowledged position. One cannot but admit that the process of arriving at this practical policy has been slow, but its consummation had to await a favourable economic situation. Between the decentralized form of orchestral policy symbolized by the Association of Symphony Orchestras and the centralized policy pursued by the B.B.C. there are differences of opinion inseparable from their different methods of attacking a common problem. Compromise can be reached on some issues, and should be sought as much as possible, but it is out of the question that in an art giving so much scope for individuality as orchestral playing compromise can be generally satisfactory. Old ideas about the nature of the machine will have to be reviewed, for it is at once a leveller of taste and a force for advancement of the art. Science will continue to concentrate power in the machine, and, as further electrophonic experiments bring satisfactory results, organizations exploiting the machine will improve their facilities. Against this mechanization of art there will always be artists who rebel, and rightly so, but the fact has nevertheless to be faced, that the discovery of the thermionic valve in 1913 revolutionized the practice of sound-production and distribution. The artist who opposes mechanization will therefore belong to a conservative opposition, for the ease with which a centralized organization can disseminate propaganda for good or for ill is already self-evident. Against this it is noted

REFORMATION

that the decentralized forces are overworked, and the time required to bring forward new ideas cannot be found without some loss of revenue-earning time. The furtherance of new music in a democratic state is linked with educational policy — by the nature of the problem mainly adult education, and adult education is as yet the Cinderella of the Ministry of Education. Facilities for preparing the public for important modern compositions exist in many adult educational movements of a voluntary kind, among which the Arts Clubs should in due course take their place, but all these have not a fraction of the power of propaganda exercised by the B.B.C., with its facilities for preliminary announcement and preparation by way of the microphone and printed publications, adequate rehearsal, presentation, and repetition, followed by issues of gramophone records by interested companies. The production of Bela Bartok's *Violin Concerto* by the B.B.C. in 1945 will serve as an example of how such an authority can direct all forms of propaganda towards a single composition.

The danger of centralization is that such forces can be used to foster an undeserving work, and the secrecy with which monopolies arrive at decisions tends to make them suspect. Freedom to propound contrary views is necessary and is not likely to be denied in this country; the support of local government authorities to their own schemes of musical enterprise is therefore in keeping with the spirit of Britain. Independent artistic enterprises also are by no means effete. The launching of Benjamin Britten's opera *Peter Grimes* at Sadler's Wells in 1945 was a splendid piece of work, to which the activities of the B.B.C. were supplementary. The history of orchestral music in the twentieth century is the story of an interplay of forces under official control and those officially free; a replica in art of the forces that are shaping the political activity of the century. Changes may be brought about more rapidly than in the past, and it is likely that they will be more confusing to the general listener, but the public has grown more intelligent in its outlook than it was in the past. At the end of the eighteenth century an

THE NATIONALIST PERIOD

influential group of French composers, including Gretry, Mehul, and Cherubini, was not ashamed to declare that 'Harmony to-day is complicated to the last degree. Singers and instruments have exceeded their natural compass. The rapidity of execution makes our music inappreciable by the ear, and one step more will plunge us into chaos.' Yet we, with the certainty of far more revolutionary changes to come in our time than ever Beethoven brought, look forward to the future with interest.

INDEX

Abel, Karl Friedrich, 74, 76 et seq., 79, 92
Academy of Antient Music, 26, 62
Albert Hall, Royal, 285
Albert, Prince Consort, 151, 189
Ancient Concerts (See Concert of Antient Music)
Arne, Dr. Thomas A., 34, 61
Arnold, Dr. Samuel, 50
Athenaeum, The, 184
Attwood, Thomas, 110, 116

Bach, C. P. E., 76, 77
Bach, J. C., 29, 34, 69, 74 et seq.
 Symphonies, 77, 79, 92
Bach, J. S., 29, 37, 95, 179, 204
Bache, Walter, 218
Banister, John, 21-3, 26, 30, 81, 190
Bantock, Sir Granville, 216, 234 et seq., 269
Bassoon, 34, 58, 145
B.B.C., 275, 278 et seq., 287, 291
 Symphony Orchestra, 272, 279 et seq., 290
Beckford, Peter, 113
Beecham Orchestra, 266, 276
Beecham, Sir Thomas, 241, 255, 257, 259 et seq., 269, 270, 276, 281, 283
Beethoven, 106, 109, 120 et seq., 187, 192
Bennett, Joseph, 178
Bennett, Sterndale, 204
Berlin Philharmonic Orchestra, 17, 278
Berlioz, 166, 181, 182, 198
Birmingham, 269 et seq.
 City Orchestra, 272 et seq.
Bishop, Sir Henry R., 178
Bottesini, 181
Boult, Sir Adrian C., 272 et seq., 283
Bournemouth, 230 et seq.
Boyce, William, 32, 60, 61, 66, 67
 Symphonies, 67 et seq.
Brahms, 212, 218, 219, 221
Brandenburg Concertos, Bach, 29
Brass, 36 et seq.
 Bands, 168, 169
Brian, W. Havergal, 249, 250

British Symphony Orchestra, 273
British Women's Symphony Orchestra, 273
Britten, Benjamin, 291
Britton, Thomas, 23-6, 45
Brodsky, Dr. Adolph, 226, 227
Buononcini, 62
Burlington, Lord, 44, 46, 47
Burney, Dr. Charles, 31, 60, 61, 68, 74, 88, 95
Byrne, W., 239

Cadenza, 57
Camm, John B., 233
Carl Rosa Opera Co., 215, 264
Cathcart, Dr. G. C., 244
Catterall Quartet, 277, 278
Cello, 42, 58
Chamberlain, Neville, 271
Chandos, Duke of, 44, 45
Charity Concerts, 62, 65, 67
Charles II, King of Britain, 20, 21, 25, 94
Chartism, 195
Cherubini, 109, 110, 127
Chopin, 193
Chorley, H. F., 149, 173, 184
Chrysander, 49
Clarinet, 33, 34, 38, 80, 96, 145, 146
Clementi, Muzio, 108, 109, 112, 113 et seq., 147, 157
Coates, Albert, 278
Cohen, Louis, 287
Concert-halls, 41
Concert of Antient Music, 78, 79, 104, 105, 109, 149
Concerto Grosso, 29, 38, 54, 55, 97, 152
Conducting, 39, 40, 67, 89, 107, 119, 254, 260
Continuo, 32, 54, 55, 59, 89, 108
Cor Anglais, 219
Cornelys, Mrs., 74, 76
Cornet-à-pistons, 162
Cornett, 33
Costa, Sir Michael, 150, 159 et seq., 171 et seq., 183, 204, 206 et seq., 239
Cowen, Sir F. H., 216, 236

INDEX

Cramer, François, 148, 149
Cramer, Jean Baptist, 106, 112 et seq., 119
Cramer, Wilhelm, 79-81
Crystal Palace, 28, 203 et seq.
Cusins, Sir W. C., 214

Daily Telegraph, 213
Davison, J. W., 183, 184, 203, 209
Debussy, 228
Defesch, William, 64
Delany, Mrs., 47
Delius, Frederick, 228, 254, 267
Deputy System, 199, 215, 248, 249
Der Freischütz, Weber, 38, 145
Diaghileff, Serge P., 267
Dibdin, Charles, 26
Dragonetti, Domenico, 134, 181

EDINBURGH, 209
Education in Music, 239, 274, 286
Elgar, Sir Edward, 228, 229, 236, 237, 254, 255, 257
Ella, John, 190, 191, 209

FESTIVAL, THE LONDON WAGNER, 221
Festivals, Provincial, 197
Field, John, 113, 114, 147
Financial Responsibilities of Orchestras, 253, 272, 283, 291
Flute, 31, 35, 50
  Bass, 59
Forster, William, 82
Forsyth Brothers, 196
Foster, Myles Birket, 107
Franck, César, 228

GALLINI, 84, 85, 92
George I, King of Britain, 47
  III, King of Britain, 78
  IV, King of Britain, 126, 127
German, Sir Edward, 234
Glasgow, 288
Goddard, Arabella, 190
Godfrey, Sir Dan E., 231 et seq., 243, 288
Great Exhibition of 1851, 167, 171, 215
Greene, Dr. Maurice, 26, 61
Grove, Sir George, 171, 205, 210 et seq.
Grüneison, C. L., 184
Gye, Frederick, 166, 169

HALLÉ ORCHESTRA, 195 et seq., 255, 276, 281, 287, 288
Hallé, Sir Charles, 147, 190 et seq., 211, 226
Handel, 24, 37-9, 41, 44 et seq., 65, 66, 71, 74, 109, 149
  Concertos, 52 et seq.
  Fireworks Music, 28, 41, 51
  Water Music, 46 et seq.
Handel's Orchestra, 50, 51
Hanover Square Rooms, 76, 85, 174, 204, 208
Harmonicon, 134
Harp, 59, 146
Harpsichord, 39, 75, 91, 113
Haweis, Rev. H. R., 172
Hawkins, Sir John, 31, 47, 74
Haydn, Joseph, 43, 45, 71, 75, 80 et seq.
  Symphonies, 92 et seq., 103, 109, 176
Hogarth, George, 61, 104, 107, 117, 140, 160, 206-8, 255
Holst, Gustav, 272, 273
Horn, 31, 32, 38, 40, 50, 59, 145
Howes, Frank, 91, 284
Hummel, J. N., 147, 157
Humphreys, Pelham, 21, 22
Hylton, Jack, 283

ILLUSTRATED LONDON NEWS, 184, 186
International Celebrity Subscription Concerts, 278

JULLIEN, LOUIS ANTOINE, 162 et seq., 208, 247

KEMBLE, CHARLES, 144
Kielmansegg, Baron, 47
Kimpton, Miss Gwynne, 274
King's Band, 21, 22, 36, 37, 71, 78

LAMBERT, CONSTANT, 68, 69
Lawes, Henry, 18, 22, 89, 90
Leader, 90, 108, 116, 176
Levien, J. Newburn, 285
Liszt, Franz, 147, 151
Liverpool, 197, 234, 262 et seq., 287
Llandudno, 230
London:
  Musical Taste, 95, 98, 104, 105
  Opera, 104, 150, 161
  Philharmonic Orchestra, 282 et seq.

INDEX

London: *cont.*
  Symphony Orchestra, 252 *et seq.*, 281
  Theatres, 160 *et seq.*
Lucas, Charles, 166
Lully, J. B., 30
Lute, 18 *et seq.*

MACE, THOMAS, 33, 40, 42, 43
Macfarren, Sir G., 189, 214
Mackenzie, Sir A. C., 216, 237
Maelzel, 123
Mainzer, Joseph, 194
Manchester, 192 *et seq.*, 252
  Brand Lane Concerts, 252 (*See also* Hallé Orchestra)
  Cecilia Society, 194
  Gentlemen's Concerts, 193-4
  Music Library, 227
Mannheim, 79, 80
Manns, Sir August, 171, 205, 210 *et seq.*, 247
Mellon, Alfred, 214
Mendelssohn, 109, 148 *et seq.*, 183, 186
Mengelberg, Willem, 278
Merseyside Orchestra, 287
Methodism, 72
Monday Popular Concerts, 190, 209
Montagu, Lady Mary Wortley, 38
Monteverdi, Claudio, 59
Morning Post, 184
Morrow, Walter, 224
Moscheles, 137
Mozart, Leopold, 64, 65
Mozart, W. A., 75, 80, 94, 109, 110, 112, 118, 145
Municipal Music, 238, 241, 270, 271, 272, 287
Musard, Philippe, 162
Musical Culture Ltd., 283
Musical Union, 190
Musicians' Union, 289

NATIONAL ASSOCIATION OF SYMPHONY ORCHESTRAS, 289
Nationalism:
  Britain, 215, 217, 237, 244
  Germany, 121, 142, 217
  Russia, 255, 267
Neate, Charles, 106, 110, 124, 129, 138
New Brighton, 234 *et seq.*

Newman, Ernest, 282
Newman, Robert, 244
New Philharmonic Society, 180, 181
New Symphony Orchestra (later the Royal Albert Hall Orchestra) 253, 265, 278
Nikisch, Arthur, 254, 255

OATES, DR. J. P., 167
Obbligato, 37
Oboe, 32-4, 54
Orchestra, constitution of, 39, 50, 51, 78, 85, 86, 96, 110, 111, 174, 180, 211, 290
Orchestral Players, status of, 62-4, 112, 150, 151, 156, 172, 198, 206, 253, 271, 276, 277, 286
Overture, 29, 68
Oxford, 63

PARIS, CONCERT SPIRITUEL, 23, 85
  Court of, 25
Parry, Sir Hubert, 216, 217, 236
Payne, Arthur, 230
Pepusch, Dr. J. C., 23, 26
Pepys, Samuel, 21, 22, 27, 94
Peri, Jacopo, 59
Philharmonic Arts Club, 289
Philharmonic Relations:
  with Beethoven, 122 *et seq.*
  with Mendelssohn, 147 *et seq.*
  with Wagner, 182 *et seq.*
Philharmonic Society (Dr. Greene's), 62
Philharmonic Society of London (now the Royal Philharmonic Society), 105 *et seq.*, 255 *et seq.*
Pianoforte, 92, 113, 147
Pitch, 239, 244, 245
Potter, Cipriani, 118, 165
Powell, Lionel, 278
Prague Philharmonic Orchestra, 17
Priestley, J. B., 283
Professional Concert, 74, 79, 80, 81, 105, 109, 118
Promenade Concerts:
  Musard, 162
  Jullien, 163 *et seq.*
  Valentino, 162
  Wood, 244 *et seq.*, 285
Proms. Circle, 289
Purcell, Henry, 22, 36, 38, 42, 43, 65, 66

295

INDEX

QUEEN'S HALL, 285
  Audiences, 251, 285

RANELAGH, 27
Ravenscroft, John, 30
Reed, W. H., 254
Reeves, Sims, 166, 190
Reynolds, John, 164
Richter, Hans, 212, 213, 219 *et seq.*,
  239, 243, 254, 257, 260
Ries, Ferdinand, 131, 135, 136
Ripieno, 55, 97, 152
Ritornello, 69
Rivière, 230, 245
Rodewald, Alfred E., 237, 262, 263, 264
Romantic Movement:
  Britain, 97 *et seq.*
  Germany, 142 *et seq.*
Rossini, 136
Royal Academy of Music, 118, 190
Royal Albert Hall Orchestra (*See* New
  Symphony Orchestra)
Royal College of Music, 215
Royal Italian Opera, 206, 207
Royal Society of Musicians, 62, 63
Russell, Thomas, 225, 283
Russian Ballet, 267
Ryan, Desmond L., 184

SACKBUT, 37
Sacred Harmonic Society, 159, 171
Safonoff, V. I., 255
St. James's Hall, 190, 209, 215
St. Paul's Cathedral, 53
Salomon, Johann Peter, 80, 81, 84 *et seq.*
  his Orchestra, 86, 110, 114
Sargeant Trumpeter, 37
Sargent, Dr. Malcolm, 274
Sax, Adolph, 167
Saxhorn, 167
Saxophone, 167
Schubert, 205, 212
Schumann, 204, 212, 227
Scottish Orchestra, 277, 288, 289
Serpent, 33, 220
Shore, Bernard, 225
Shore, John, 36, 37
Sibelius, Jean, 269
Sinfonia avanti l'opera, 29, 68
Smart, Sir George, 86, 116, 118, 133, 137, 144, 208, 220
Snow, Valentine, 37

Sons of the Clergy, Festival of, 65, 67
Spohr, Dr. Louis, 109, 119, 151, 152
Stamitz, 79, 80, 106
Standard, The, 184
Stanford, Sir Charles V., 216, 217, 236
Strauss, Ludwig, 197
Strauss, Richard, 256, 257, 267
Stravinsky, Igor, 268
Strings, 36, 40, 42
Subsidies, Municipal, 240, 271, 287, 288
Sullivan, Sir Arthur, 214, 223
Sunday Times, 186
Symphonies:
  of J. C. Bach, 77, 58
  of Boyce, 69 *et seq.*
  of Haydn, 96, 97
Symphony, 29, 67 *et seq.*

THALBERG, 214
Theatres, 160 *et seq.*
Three Choirs Festival, 60, 66, 252
Times, The, 183, 203
Tone-colour, 36, 58, 59
Torquay, 238
Trombone, 59 (*See also* Sackbut)
Trumpet, 31, 32, 36-8, 50, 51, 71

VALENTINO, HENRI JUSTIN ARMAND, 162, 163
Vauxhall Gardens, 27, 28, 51, 76
Vienna, 120, 121, 132
Vienna Philharmonic Orchestra, 17
Viola, 42, 145
Violin, 19 *et seq.*, 40, 44
Viols, 18 *et seq.*, 40, 94
Viotti, 113 *et seq.*

WAGNER, RICHARD, 156, 182, 183 *et seq.*, 218
Waits, 30, 32, 36
Walsh, 48-50
Weber, 38, 142 *et seq.*
Weideman, 62, 63
Wilhelm II, Emperor of Germany, 243
Wireless Telephony, 275
Wood, Anthony, 18-20, 23, 30, 94
Wood, Sir Henry J., 227, 230, 238, 244 *et seq.*, 268
Wood-wind, 31, 34, 36, 40, 91, 97

YBARRONDO, DE, 237